T0140440

Studies in Big Data

Volume 79

Series Editor

Janusz Kacprzyk, Polish Academy of Sciences, Warsaw, Poland

The series "Studies in Big Data" (SBD) publishes new developments and advances in the various areas of Big Data- quickly and with a high quality. The intent is to cover the theory, research, development, and applications of Big Data, as embedded in the fields of engineering, computer science, physics, economics and life sciences. The books of the series refer to the analysis and understanding of large, complex, and/or distributed data sets generated from recent digital sources coming from sensors or other physical instruments as well as simulations, crowd sourcing, social networks or other internet transactions, such as emails or video click streams and other. The series contains monographs, lecture notes and edited volumes in Big Data spanning the areas of computational intelligence including neural networks, evolutionary computation, soft computing, fuzzy systems, as well as artificial intelligence, data mining, modern statistics and Operations research, as well as self-organizing systems. Of particular value to both the contributors and the readership are the short publication timeframe and the world-wide distribution, which enable both wide and rapid dissemination of research output.

** Indexing: The books of this series are submitted to ISI Web of Science, DBLP, Ulrichs, MathSciNet, Current Mathematical Publications, Mathematical Reviews, Zentralblatt Math: MetaPress and Springerlink.

More information about this series at http://www.springer.com/series/11970

Raghvendra Kumar · Rohit Sharma ·
Prasant Kumar Pattnaik
Editors

Multimedia Technologies in the Internet of Things Environment

Editors
Raghvendra Kumar
Department of Computer Science
and Engineering
GIET University
Gunupur, Odisha, India

Rohit Sharma
Department of Electronics
and Communication Engineering
SRM Institute of Science and Technology
Ghaziabad, Uttar Pradesh, India

Prasant Kumar Pattnaik
School of Computer Engineering
KIIT Deemed to be University
Bhubaneswar, Odisha, India

ISSN 2197-6503 ISSN 2197-6511 (electronic)
Studies in Big Data
ISBN 978-981-15-7967-7 ISBN 978-981-15-7965-3 (eBook)
https://doi.org/10.1007/978-981-15-7965-3

This Springer imprint is published by the registered company Springer Nature Singapore Pte Ltd.
The registered company address is: 152 Beach Road, #21-01/04 Gateway East, Singapore 189721, Singapore

Preface

The main objective of this book publication is to explore the concepts of Internet of things, and biomedical and cyber-physical systems along with the recent research and development. It also includes various real-time applications and case studies in the field of engineering and technologies used. As populations grow and resources become scarcer, the efficient usage of these limited goods becomes more important.

Chapter "Smart Control and Monitoring of Irrigation System Using Internet of Things" aims to control the agriculture water pump motor by using IoT-based controller, by sensing the water flow and running time of motor and by analyzing the status of motor. Our concept is to monitor and control the motor and also reduce some difficulty of ON/OFF.

Chapter "Blockchain-Based Cyberthreat Mitigation Systems for Smart Vehicles and Industrial Automation" discusses various use cases of blockchain in smart vehicles and ongoing advancement from motorized industries in addition to educational organizations. Blockchain in smart vehicles functions as a public distributed ledger which is capable of recording dealings between two groups, though the acceptance and execution of blockchain technologies in numerous organizations and facilities is a difficult job. The goal of this paper is to classify the obstacles for the effective execution of blockchain technologies in the companies.

Chapter "IT Convergence-Related Security Challenges for Internet of Things and Big Data" demonstrates the Internet-of-things- and big data-related applications and analyzes various security issues whenever the environment is diagnosed. The latest security issues may happen in every element in the Internet of things because an element is communicated to the security vulnerability. Whenever the data is generated through the big data and the Internet of things surroundings, these kinds of problems are identified according to the security-related relationship. The necessity of the security parameters of the ICT-based environment is analyzed in a detailed way.

Chapter "Applicability of Industrial IoT in Diversified Sectors: Evolution, Applications and Challenges" presents an overview of different emerging applications of IoT such as blockchain applications in Industry 4.0 and IIoT settings. As security is considered the main aspect in any model, therefore security metrics have been highlighted along with IoT-based architecture. Later section of the paper describes evolution of IIoT, revolution in industrial sector, collaboration of Industry 4.0 and IIoT, industrial automation, cyber-security and data analytics, blockchain enrolled with IIoT, applications in public domain, etc.

Chapter "Recent Emerging Technologies for Intelligent Learning and Analytics in Big Data" gives an overview of recent emerging technologies for intelligent learning and analytics in multimedia and at the same time for online streaming processing of multimedia data for applications and system layers in urban life, particularly in education. These technologies are designed to help selection and usage of the right blend of multiple technologies for large volumes of data according to their needs and the requirements of concrete apps.

Chapter "Real-Time Health System (RTHS) Centered Internet of Things (IoT) in Healthcare Industry: Benefits, Use Cases and Advancements in 2020" discusses the level of medical devices and healthcare assets, the people level and the non-medical asset level (e.g., hospital building assets). Furthermore, we have stated a few drivers and benefits that influence the evolutions and forecasts in the implementation of IoT in health care. Most of them fit in overall healthcare drivers such as aging populations, the changing behavior and demands of patients (consumerization of health care and patient centricity) and of healthcare workers, budgetary challenges and the improvement of care quality.

Chapter "Building Intelligent Integrated Development Environment for IoT in the Context of Statistical Modeling for Software Source Code" is the study of intelligent IDEs extended to IoT environment context. To begin with, we focus our study on understanding the IDEs and their capability. Exploration will focus on open-source IDE for mobile, and we look for better understanding the landscape so that we can extend the learning to the IoT world. In this work, building low-cost IDE for mobile is focused open. Also, interestingly, exploration focuses on open-source components and possibility of putting them together. Then in the next part, we explore the IDE for Internet of things (IoT), with focus on open-source ecosystem. In this review, we extend the exploration of IoT into device management, data management, communication, intelligent data processing, security and privacy and application deployment areas. This gives a greater insight into the IoT world to extend the need of intelligent IDEs to IoT world. Then to get the context of IDE in the machine learning context, we explore the topic of building optimal IDE for feature engineering which is one of the key phases in machine learning life cycle. Since machine learning projects are highly data-oriented ecosystem, learning of IDE and its insights in this area will provide rich insights into the main theme of discussion which is intelligent IDE for IoT ecosystem.

Chapter "Visualization of COVID-19 Pandemic: An Analysis Through Machine Intelligent Technique Toward Big Data Paradigm" discusses a real-time data set from World Health Organization (WHO) which is collected and analyzed. The nature of the data set is as the concept of big data. A statistical analysis is performed on the data set and produces the results. The data for India is only considered for analysis. The number of death cases and confirmed cases in India during the period of 15 weeks from 28 February 2020 to 7 June 2020. Finally, the analysis reported that in India rate of death cases is less than rate of cure cases.

Chapter "Multimedia Security and Privacy on Real-Time Behavioral Monitoring in Machine Learning IoT Application Using Big Data Analytics" targets two zones; from the start, the paper gives an examination of the IoT layered engineering to set up an IoT space with the security challenges/attack. From that point on, the paper proposes a response that can protect the security of mixed media information within an Internet of Multimedia Things (IoMT) state at its observation layer. The proposed security structure would improve the mixed media to operate adequately while at the same time safeguarding the privacy of the transmitted data and protecting people.

Chapter "A Robust Approach with Text Analytics for Bengali Digit Recognition Using Machine Learning" aims to classify the Bengali numerals obtained from the NumtaDB data set with the implementation of a convolutional neural network (CNN). Moreover, we have also conducted a survey based on Bengali numerals and performed a sentiment analysis on the recorded responses of our subjects using the same classification technique and thereafter have touched upon an area of application of vernacular digit recognition which has seldom gone unnoticed. We have also endeavored to provide an in-depth literature survey of the works that have been done in this area previously. Our classification model gives us a whopping accuracy of 98.40%.

Chapter "Internet of Things-Based Security Model and Solutions for Educational Systems" identifies challenges in security issues in IoT-based educational systems and some probable solutions on security. In this research, the authors propose the incremental Gaussian mixture model (IGMM), blockchain, EdgeSec as a probable solution for security and machine learning (ML) techniques. In the proposed model, we discussed the solutions for IGMM for authorizing the device, Blockchain for the encryption of data during transfer in the information network, ML algorithms for identifying and authorizing devices, EdgeSec offers a security profile to collect a huge amount of data about each device in the connected IoT environment. The identified model is used for enhancing security in IoT-based educational systems.

The aim of this book is to support the computational studies at the research and post-graduation level with open problem-solving technique. We are confident that it will bridge the gap for them by supporting novel solution to support in their problem solving. At the end, the editors have taken utmost care while finalizing the chapter to the book, but we are open to receive your constructive feedback, which will enable us to carry out necessary points in our forthcoming books.

Gunupur, India	Raghvendra Kumar
Ghaziabad, India	Rohit Sharma
Bhubaneswar, India	Prasant Kumar Pattnaik

About This Book

Multimedia communication in the Internet of things (IoT) can potentially reach into a vast array of areas and touch people's lives in profound and different ways. A vast amount of research work has been carried out in this domain, targeting different aspects of data analytics. To serve this purpose, it is proposed to conduct a comprehensive overview of the state-of-the-art research work on multimedia analysis in IoT applications. It also aims to bridge the gap between multimedia concept and solutions by providing the current IoT frameworks, their applications, the strengths and limitations of the existing methods, and the future directions.

Key Features

1. The proposed book addresses the complete functional framework in the area of multimedia data, IoT and smart computing techniques.
2. The proposed book conducts a comprehensive overview of the state-of-the-art research work on multimedia analysis in IoT applications.
3. The proposed book aims to bridge the gap between multimedia concepts and solutions by providing the current IoT frameworks, their applications in multimedia analysis, the strengths and limitations of the existing methods, and the future directions in multimedia IoT analytics.

Contents

Editors and Contributors

About the Editors

Dr. Raghvendra Kumar is working as Associate Professor in Computer Science and Engineering Department at GIET University, India. He received B. Tech, M.Tech and Ph.D. in Computer Science and Engineering, India, and Postdoc Fellow from Institute of Information Technology, Virtual Reality and Multimedia, Vietnam. He serves as Series Editor Internet of Everything (IOE): Security and Privacy Paradigm, Green Engineering and Technology: Concepts and Applications, publishes by CRC press, Taylor & Francis Group, USA, and Bio-Medical Engineering: Techniques and Applications, Publishes by Apple Academic Press, CRC Press, Taylor & Francis Group, USA. He also serves as acquisition editor for Computer Science by Apple Academic Press, CRC Press, Taylor & Francis Group, USA. He has published number of research papers in international journal (SCI/SCIE/ESCI/Scopus) and conferences including IEEE and Springer as well as serve as organizing chair (RICE-2019, 2020), volume Editor (RICE-2018), Keynote speaker, session chair, Co-chair, publicity chair, publication chair, advisory board, Technical program Committee members in many international and national conferences and serve as guest editors in many special issues from reputed journals (Indexed By: Scopus, ESCI, SCI). He also published 13 chapters in edited book published by IGI Global, Springer and Elsevier. His researches areas are Computer Networks, Data Mining, cloud computing and Secure Multiparty Computations, Theory of Computer Science and Design of Algorithms. He authored and Edited 23 computer science books in field of Internet of Things, Data Mining, Biomedical Engineering, Big Data, Robotics, and IGI Global Publication, USA, IOS Press Netherland, Springer, Elsevier, CRC Press, USA.

Rohit Sharma is working as an Assistant Professor in the Department of Electronics and Communication Engineering, SRM Institute of Science and Technology, Delhi NCR Campus Ghaziabad, India. He is an active member of ISTE, IEEE, ICS, IAENG, and IACSIT. He is an editorial board member and reviewer of more than 12 international journals and conferences, including the topmost journal IEEE Access and IEEE Internet of Things Journal. He serves as a Book Editor for 7 different titles to be published by CRC Press, Taylor & Francis Group, USA and Apple Academic Press, CRC Press, Taylor & Francis Group, USA, Springer, etc. He has received the Young Researcher Award in "2nd Global Outreach Research and Education Summit & Awards 2019" hosted by Global Outreach Research & Education Association (GOREA). He is serving as an Editor and Organizing Chair to 3rd Springer International Conference on Microelectronics and Telecommunication (2019), and have served as the Editor and Organizing Chair to 2nd IEEE International Conference on Microelectronics and Telecommunication (2018), Editor and Organizing Chair to IEEE International Conference on Microelectronics and Telecommunication (ICMETE-2016) held in India, Technical Committee member in "CSMA2017, Wuhan, Hubei, China", "EEWC 2017, Tianjin, China" IWMSE2017 "Guangzhou, Guangdong, China", "ICG2016, Guangzhou, Guangdong, China" "ICCEIS2016 Dalian Liaoning Province, China".

Prasant Kumar Pattnaik Ph.D (Computer Science), Fellow IETE, Senior Member IEEE is a Professor at the School of Computer Engineering, KIIT Deemed University, Bhubaneswar. He has more than a decade of teaching and research experience. Dr. Pattnaik has published numbers of Research Papers in peer-reviewed International Journals and Conferences. He also published many edited book volumes in Springer and IGI Global Publication. His areas of interest include Mobile Computing, Cloud Computing, Cyber Security, Intelligent Systems and Brain Computer Interface. He is one of the Associate Editor of Journal of Intelligent & Fuzzy Systems, IOS Press and Intelligent Systems Book Series Editor of CRC Press, Taylor Francis Group.

Contributors

Raghavendra Rao Althar QMS, First American India, Bangalore, India

C. Anuradha Department of Computer Science and Engineering, SCAD College of Engineering and Technology, Cheranmahadevi, India

G. Arun Sampaul Thomas CSE Department, J.B. Institute of Engineering and Technology, Hyderabad, India

Valentina E. Balas Automatics and Applied Informatics, Aurel Vlaicu University of Arad, Arad, Romania

Bharat Bhushan School of Engineering and Technology, Sharda University, Noida, India

Korhan Cengiz Faculty of Engineering, Trakya University, Edirne, Turkey

K. Elangovan Department of Electronics and Communication Engineering, Sriram Engineering College, Chennai, Tamil Nadu, India

R. Ganesh Babu Department of Electronics and Communication Engineering, SRM TRP Engineering College, Tiruchirappalli, Tamil Nadu, India

Bhavya Gaur Business Technology Analyst, ZS Associates, Pune, India

T. Gopu Department of Electronics and Communication Engineering, KLN College of Engineering, Sivagangai, India

Saptarshi Gupta Electronics and Communication Engineering, SRM Institute of Science & Technology, Delhi-NCR, India

R. Jeyapandiprathap Department of Electronics and Communication Engineering, KLN College of Engineering, Sivagangai, India

T. Kalimuthu Department of Electronics and Communication Engineering, St. Mother Theresa Engineering College, Thootukudi, India

P. Karthika Department of Computer Applications, Kalasalingam Academy of Research and Education, Srivilliputhur, Tamil Nadu, India

Ila Kaushik Krishna Institute of Engineering and Technology, Ghaziabad, U.P., India

Kottilingam Kottursamy SRM Institute of Science and Technology, Kattankulathur Campus, Chennai, India

Raghvendra Kumar Department of Computer Science and Engineering, GIET University, Gunupur, India

Gunjan Madaan HMR Institute of Technology and Management (1) School of Engineering and Technology, Sharda University, India (2) and Department of Computer Science and Engineering, GIET University, Gunupur, India (3), New Delhi, India

Sudhanshu Maurya School of Computing, Graphic Era Hill University, Dehradun, Uttarakhand, India

Proshikshya Mukherjee School of Electronics Engineering, KIIT DU, Bhubaneswar, India

Basak Ozyurt Faculty of Applied Sciences, Trakya University, Edirne, Turkey

M. Palpandi Department of Electronics and Communication Engineering, KLN College of Engineering, Sivagangai, India

Ranjit Patnaik Department of Computer Science and Engineering, GIET University, Gunupur, India

Prasant Kumar Pattnaik School of Computer Engineering, KIIT DU, Bhubaneswar, India

Dipam Paul School of Electronics Engineering, KIIT DU, Bhubaneswar, India

P. Rajkumar Department of Electronics and Communication Engineering, KLN College of Engineering, Sivagangai, India

Y. Reeginal Department of Physics, SCAD College of Engineering and Technology, Cheranmahadevi, India

Y. Harold Robinson School of Information Technology and Engineering, Vellore Institute of Technology, Vellore, India

Debabrata Samanta Department of Computer Science, CHRIST (Deemed to be University), Bangalore, India

Manash Sarkar Computer Science & Engineering, SRM Institute of Science & Technology, Delhi-NCR, India

Rohan Sethi HMR Institute of Technology and Management, New Delhi, India

Nikhil Sharma HMR Institute of Technology and Management, New Delhi, India

Rohit Sharma SRM Institute of Science and Technology, SRM University, Ghaziabad, India

Krishna Kant Singh KIET Group of Institutions, Ghaziabad, India

K. Sivakrishna Department of Computer Science and Engineering, GIET University, Gunupur, India

K. Srujan Raju Department of Computer Science and Engineering, CMR Technical Campus, Hyderabad, India

Tuna Topac Department of Computational Sciences, Trakya University, Edirne, Turkey

C. Vengatesh Department of Electronics and Communication Engineering, KLN College of Engineering, Sivagangai, India

G. Vijayarani Department of Computer Science and Engineering, SCAD College of Engineering and Technology, Cheranmahadevi, India

Smart Control and Monitoring of Irrigation System Using Internet of Things

P. Rajkumar, C. Vengatesh, M. Palpandi, T. Gopu, R. Jeyapandiprathap, T. Kalimuthu, and Y. Harold Robinson

Abstract The main objective of this paper is to reduce human intervention and increase the irrigation efficiency by controlling and monitoring of irrigation system using IoT. Interconnection of number of devices through Internet describes the Internet of things (IoT). Every object is connected with each other through unique identifier so that data can be transferred without human to human interaction. The objective of project is to control the agriculture water pump motor by using IOT-based controller, by sensing the water flow and running time of motor and by analyzing statues of motor. The time of situation water problem occurs in the agriculture area. Then motor running time is reducing due to the waste of water in the irrigation system. Our concept is to monitor and control the motor and also reduce some difficulty to ON/OFF.

P. Rajkumar · C. Vengatesh · M. Palpandi · T. Gopu · R. Jeyapandiprathap
Department of Electronics and Communication Engineering, KLN College of Engineering, Sivagangai, India
e-mail: rajkumar180595@gmail.com

C. Vengatesh
e-mail: acvengat@gmail.com

M. Palpandi
e-mail: mtpandi555@gmail.com

T. Gopu
e-mail: gopu70@gmail.com

R. Jeyapandiprathap
e-mail: jppradath@gmail.com

T. Kalimuthu
Department of Electronics and Communication Engineering, St. Mother Theresa Engineering College, Thootukudi, India
e-mail: kalimuthu.t@gmail.com

Y. Harold Robinson (✉)
School of Information Technology and Engineering, Vellore Institute of Technology, Vellore, India
e-mail: yhrobinphd@gmail.com

R. Kumar et al. (eds.), *Multimedia Technologies in the Internet of Things Environment*, Studies in Big Data 79, https://doi.org/10.1007/978-981-15-7965-3_1

Keywords Soil moisture · Motor · Water flow of irrigation system

1 Introduction

Irrigation [1] is applying water to the land artificially. Water [2] is one of the precious resource and very important factors for farming. General problems in farming are under watering or overwatering [3]. Agriculture [4] plays major role in the economy of the country. More than 70% of Indian population relies on agriculture for their sustenance as the contribution of agriculture [5] to gross domestic product [6] is declining nowadays, we are in urge to increase crop productivity with efficient and effective water usage [7]. In agriculture, irrigation is the important factor as the monsoon rainfalls are unpredictable and uncertain. Agriculture in the face of water scarcity has been a big challenge [8]. There exists a demand for colossal technical knowledge to make irrigation systems more efficient and the proposed work as photographs (field images) taken through Web [9]. Overwatering is starting the water cycle too early and running it for longer period than what it is necessary; by doing this practice, the crop will be damaged and production reduces [10]. If human intervention is more than this, under and overwatering take place because of small human errors [11]. In recent years, revolutionary changes in information and communication technologies such as computers have been taking place due to the development of various new technologies such as artificial intelligence-based Internet of things [12], cloud computing [13] and big data [14] in the field of agricultural revolution process [15]. The sensor data is generated using Internet of things in various fields to promote the agriculture [16] through the automation techniques and the farmers able to know the status of the current scenario and other related improvement in the particular field [17]. The sensor data processing such as temperature [18], ultrasound, infrared, camera and vibration [19] must be analyzed quickly and accurately in each field [20] especially in the agricultural field to promote the latest techniques in the agricultural field by the farmers [21]. Because sensor data is mostly informal data, using big data enables qualitative analysis rather than general analysis for the improvements in the agriculture field [22]. The Internet of things [23] is recognized as a trend in data collection and analysis solutions used in today's diverse fields and is applied to various areas that require formal and informal data analysis [24]. The government sector has introduced several latest techniques especially the farmers from the rural are to know about the every process [25] and update in the field of agricultural sectors [26]. Unlike traditional analysis solutions [27], large amounts of data are stored in a distributed manner, so, storage space efficiency and analysis speed are different [28].

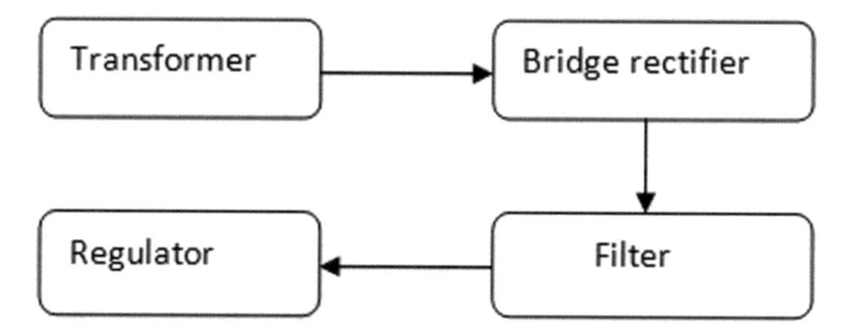

Fig. 1 Power supply module

2 Devices Used

2.1 IoT Context for Smart Irrigation

Along with various IoT application areas, agriculture includes cultivation and water management and motor monitoring. A wide range of sensors can be used for agricultural applications, soil moisture, temperature, humidity, ultraviolet and solar radiations, wind vanes, pluviometer, etc. IOT comprises deployment of sensors at the respective application fields and storage of their data cloud for processing. The same concept can be used for irrigation system for monitoring and controlling. There are various communication technologies for IoT. The IoT architecture includes sensors and actuators connected to IoT gateway via various local wireless networks. The other side of the gateway is connected to IP-based wireless actuator network and can be accessed from cloud systems. Controller algorithm is required to run on IoT gateway. The power supply of 5 v is generated using the above schematic. Step down transformer is used for 12 v supply, and bridge rectifier regulator and filter are used to produce 5 v supply which is used in the circuit. Internet of things (IoT) is widely used in connecting devices and collecting data information. Internet of things is used with IoT frameworks to handle and interact with data and information. IoT is applicable in various methodologies of agriculture (cloud technology, WSNs, GPS module, etc.,). Figure 1 demonstrates the power supply module.

2.2 Water Flow Sensor

It consists of a plastic valve body, a water rotor and a hall-effect sensor. When water flows through the rotor, rotor rolls. Its speed changes with different rate of flow. This one is suitable to detect flow in water dispenser or coffee machine in Fig. 2.

Fig. 2 Waterflow sensor

2.3 Soil Moisture

A soil moisture sensor measures the quantity of water contained in a material, such as soil on a volumetric or gravimetric basis. To obtain an accurate measurement, a soil temperature sensor is also required for calibration and illustrated in Fig. 3.

The sensors used are already discussed. Let us discuss about data acquisition from sensors one by one. The sensor is interfaced with Arduino microcontroller and programmed. Once it is programmed, it is placed inside a box and kept in the field. The soil moisture sensor has two probes which are inserted into the soil. The probes are used to pass current through the soil. The moisture soil has less resistance and hence passes more current through the soil; whereas, the dry soils have high resistance and pass less current through the soil. The resistance value helps detecting the soil moisture.

3 Proposed Methodology

The proposed system is smart objects embedded with sensors which enables interaction with the physical and logical worlds according to the concept of IoT. In this paper, proposed system is based on IoT that uses real-time input data. Smart farm irrigation system uses android phone for remote monitoring and controlling of drips through wireless sensor network. Zigbee is used for communication between sensor nodes and base station. Real-time sensed data handling and demonstration on the server are accomplished using Web-based Java graphical user interface. Wireless monitoring of field irrigation system reduces human intervention and allows remote monitoring and controlling on android phone. Cloud computing is an attractive solution to the large amount of data generated by the wireless sensor network. The whole pin description diagram is demonstrated in Fig. 4.

Fig. 3 Soil moisture

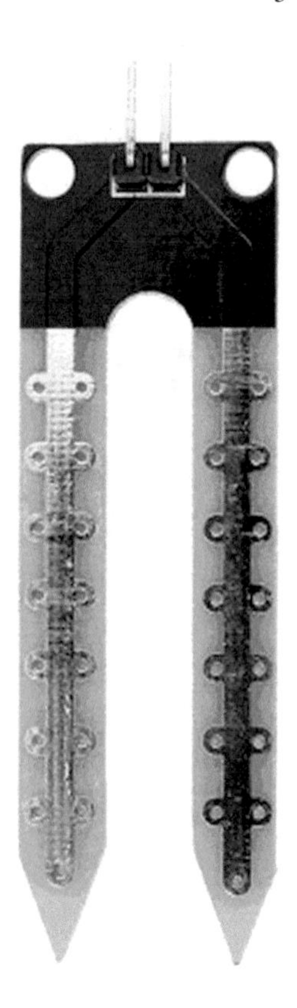

Node microcontroller unit (Node MCU) is an open-source IoT platform. It includes firmware which runs on the ESP8266 Wi-Fi SoC from express if and hardware which is based on the ESP-12 module. The term "Node MCU" by default refers to the firmware rather than the dev kits. This ESP32 Node MCU contains firmware that can run on ESP32 Wi-Fi SoC chips and hardware based on ESP-32S modules. You can create a Web server; send HTTP requests, control outputs, read inputs and interrupts, send emails, post tweets, build IoT gadgets and much more in Fig. 5.

The block diagram shown in Fig. 6 supply given to node MCU supply is separated through the soil moisture sensor, water flow sensor and motor. The moisture sensor collects data from soil water content level and sent to the node MCU. If soil water content level is less than 70%, the motor will start and then water flow rate is measured. If the level is less, then the define value will be automatically stop. The motor status and sensor status are collected and the collected data for mobile from node microcontroller unit (MCU) using Wi-Fi.

Algorithm:

Step 1:
Start the process

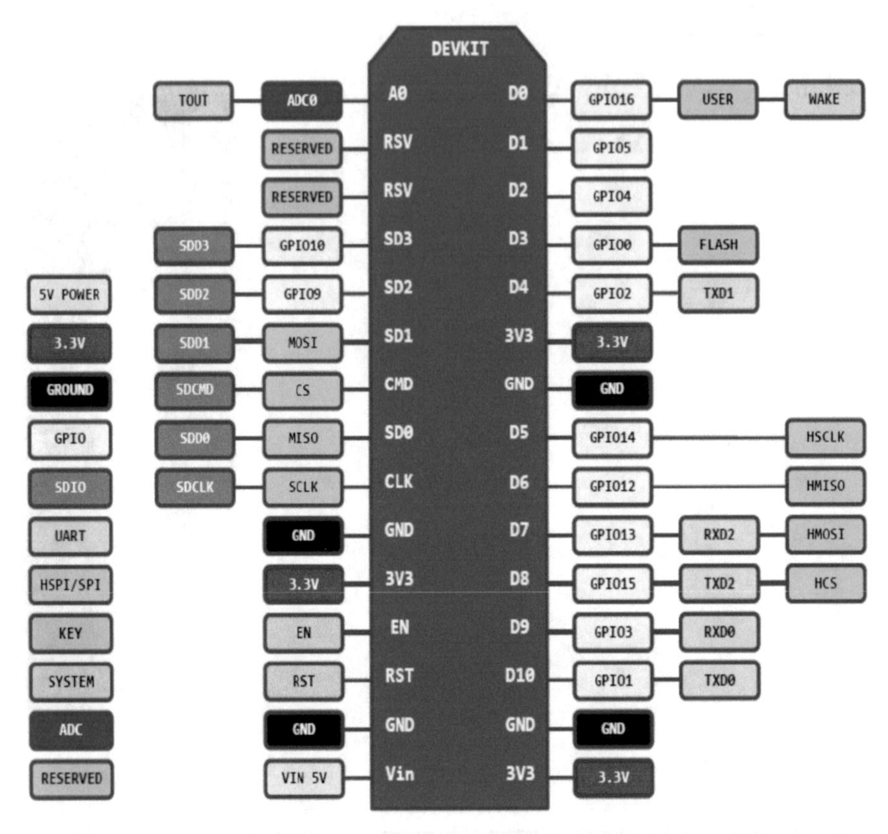

Fig. 4 Pin description diagram

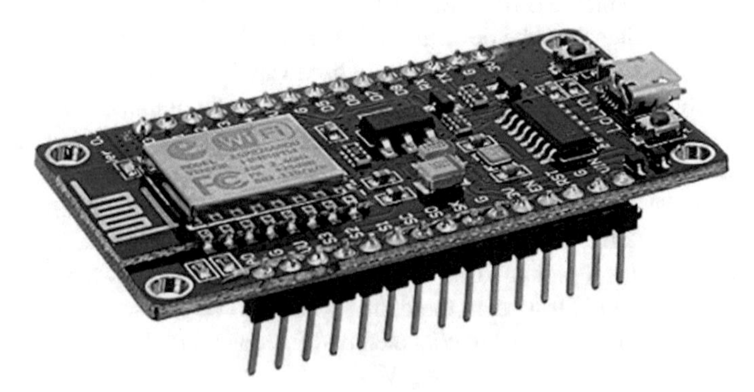

Fig. 5 Node MCU

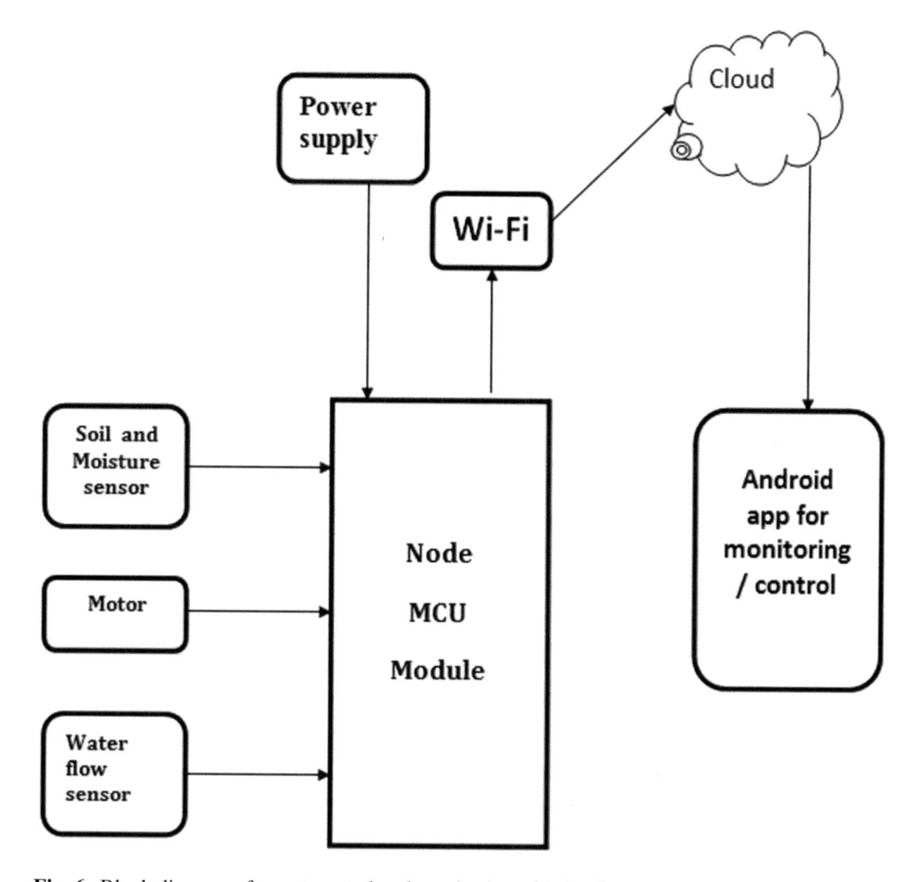

Fig. 6 Block diagram of smart control and monitoring of irrigation system using IoT

Step 2:
Check the default function

Step 3:
Check if the soil moisture is level less than 70 percentage, then only switch ON the motor, otherwise go to default function

Step 4:
If running time of the motor exceeds more than 30 min and water level is zero, then go to motor and switch OFF. Otherwise the process of app control begins

Step 5:
End of the process.

The entire process is illustrated in Fig. 7.

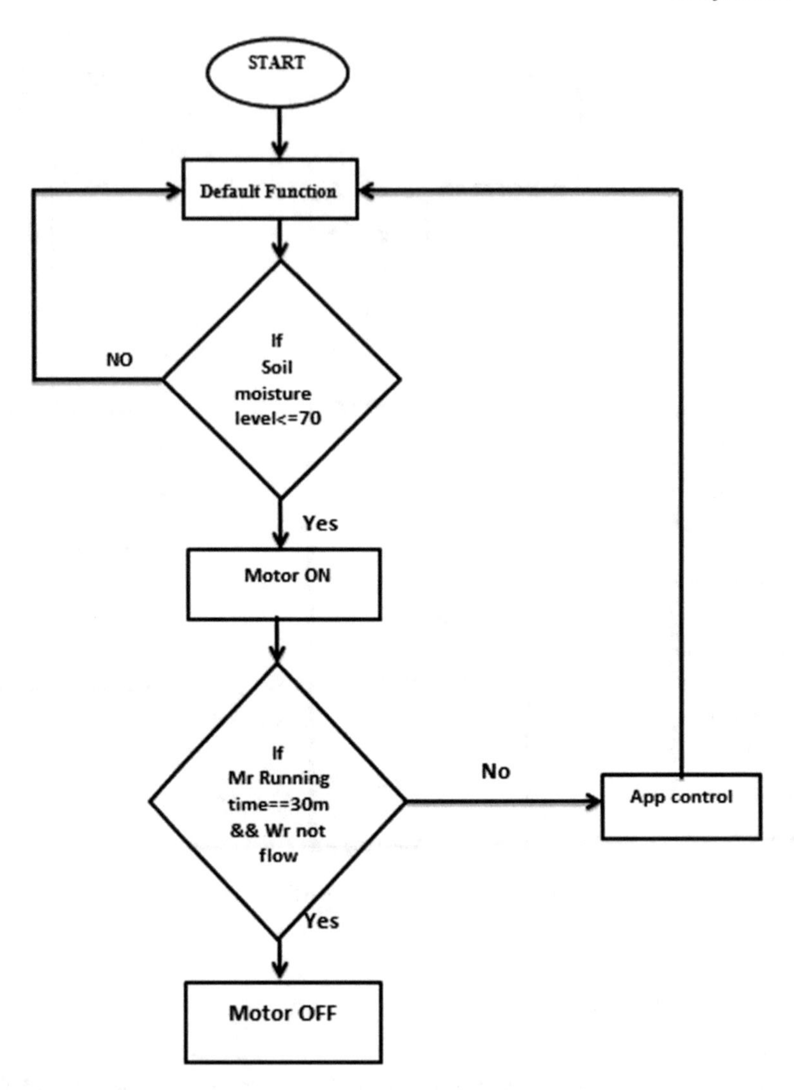

Fig. 7 Flow diagram

4 Results and Discussion

This paper proposes an idea of combining the latest technology into the agricultural field to turn the traditional methods of irrigation to modern methods, thus making it easy productive and economical. The need of this systems gives an intelligent way to provide water only to those places where it is required and in the necessary quantity, and it gives good scope in future agriculture field. The advantages like water saving and labor saving are initiated using sensors that work smart as they are programmed. This concept of modernization of agriculture is simple, affordable and operable. Table

Table 1 Specification

Range	Accuracy	Typical resolution	Power (DC)	Temperature (C)	Dimensions (cm)
0–40% water content	±4% typical	0.1%	3 mA, 5 V	−40 to + 60	8.9 * 1.8 * 0.7

Table 2 Parameters for performance

Voltage	3.3 V
Current consumption	10μA ~ 170 mA
Processor speed	80 ~ 160 MHz
RAM	32 K + 80 K
GPIOs	17
ADC	1 input with 1024 step resolution
Processor	Tensilica L106 32bit
Flash memory attachable	16 MB max (512 K normal)

1 demonstrates the specification, and Table 2 illustrates the performance parameters for this proposed approach.

Large potential of our Indian agriculture is yet untapped, and we still have miles to travel in this arena of research as we have different soil textures in different regions of our state. Farmers can be benefitted by the actual implementation of this projected program. Real challenges that were faced and that are yet to be overcome in reality are the inter-networking of the nodes in an. In future we are using a hyperspectral Camera for capture and measure the soil humility. Figure 8 demonstrates the simulation result in detail.

5 Conclusion

This chapter is about, from irrigation purpose, the irrigation system on smart uses optimal resources to improve efficiency. According to the soil parameters, the automation is achieved by turning the motor ON and OFF using the node MCU. The simulation result proved that the proposed methodology has improved accuracy rather than the related methodologies.

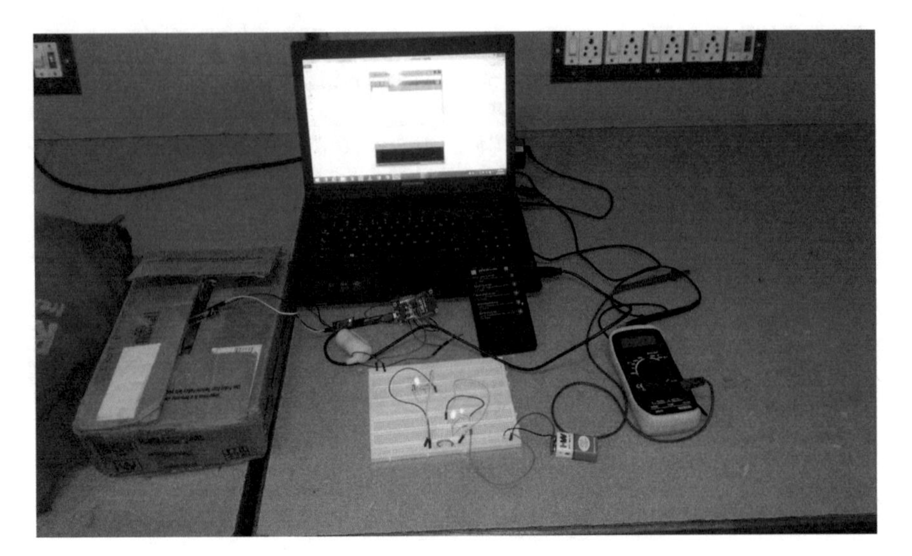

Fig. 8 Simulation result

References

1. Santhana Krishnan, R., Golden Julie, E., Harold Robinson, Y., Raja, S., Kumar, R., Thong, P. H., & Son, L. H. (2020). Fuzzy logic based smart irrigation system using internet of things. *Journal of Cleaner Production, 252*, 119902.
2. Gutierrez, J., Villa-Medina, J. F., Nieto-Garibay, A., & Porta-Gandara, M. A. (2014). Automated irrigation system using WSNs and GPRS module. *IEEE Transactions on Instrumentation and Measurement 2014.*
3. Rajpal, A., Jain, S., Khare, N., & Shukla A. K. (2011). Microcontroller based automatic irrigation system with moisture sensors. In *International Conference on Science and Engineering (ICSE 2011).*
4. Lakshminarayanan, K., Santhana Krishnan, R., Golden Julie, E., Harold Robinson, Y., Kumar, R., Son, L. H., Hung, T. X., Samui, P., Ngo, P. T. T., & Tien Bui, D. (2020). A new integrated approach based on the iterative super-resolution algorithm and expectation maximization for face hallucination. *Applied Science*, 10, 718. https://doi.org/10.3390/app10020718.
5. Stamenković, Z., Randjić, S., Santamaria, I., Pešović, U., Panić, G., & Tanasković, S. (2016). Advanced wireless sensor nodes and networks for agricultural applications. In *2016 24th Telecommunications Forum (TELFOR)*, Belgrade, pp. 1–8.
6. Sun, Y., Song, H., Jara A. J., Bie, R. (2016). Internet of things and big data analytics for smart and connected communities. In *IEEE Access, vol. 4*, pp. 766–773.
7. Balaji, S., Golden Julie, E., Harold Robinson, Y., Kumar, R., Thong, P. H., & Son, L. H. (2019). Design of a security-aware routing scheme in mobile ad-hoc network using repeated game model. *Computer Standards & Interfaces, 66.*
8. Ngangue Ndih E. D., & Cherkaoui, S. (2016). On enhancing technology coexistence in the IoT era: ZigBee and 802.11 case. In *IEEE Access, vol. 4*, 1835–1844.
9. Hai, L., He C., Tang Y., & Ping, H. S. (2012). Research and application of service-oriented scholar cloud platform. *Journal of Convergence Information Technology JCIT*, *7*(5), 333–339.
10. Santhana Krishnan, R., Golden Julie, E., Harold Robinson, Y., Kumar, R., Son, L. H., Tuan, T. A., & Long, H. V. (2019). Modified zone based intrusion detection system for security enhancement in mobile ad-hoc networks. *Wireless Networks*, 1–15.

11. Choudhary, M., Dhone, S., Jadhav, A., Dhandal, C., & Nighot, J. M. (2015). Scheduling, controlling and monitoring of agricultural devices using android application. *International Journal of Advanced Research in Computer Engineering & Technology (IJARCET)*, *4*(4).
12. Kansara, K., Zaveri, V., Shah, S., Delwadkar, S., & Jani, K. (2015). Sensor based automated irrigation system with IOT: A technical review. In K. Kansara et al (eds.), *International Journal of Computer Science and Information Technologies*, *6*(6), 5331–5333.
13. Harold Robinson, Y., Golden Julie, E., Krishnan S., Kumar, R., & Son, L. H. (2019). DRP: Dynamic routing protocol in wireless sensor networks. *Wireless Personal Communications*, 1–17 (Springer).
14. Balraj, B., & Arulmozhi, M. (2017) Isolation, characterisation and enzymatic activity of *Streptomyces* sp. and its pH control during fermentation process. *IET-Systems Biology*. ISSN: 1751-8849; https://doi.org/10.1049/ietsyb.2016.0048.
15. Kumar, A., Kamal, K., Arshad, M. O., Mathavan, S., & Vadamala, T. (2014). Smart irrigation using low-cost moisture sensors and XBee-based communication. In *IEEE Global Humanitarian Technology Conference (GHTC 2014)*, San Jose, CA, pp. 333–337.
16. Harold Robinson, Y., Santhana Krishnan, R., Golden Julie, E., Kumar, R., Son, L. H., & Thong, P. H. (2019). Neighbor knowledge-based rebroadcast algorithm for minimizing the routing overhead in mobile ad-hoc networks. *Ad Hoc Networks*, *93*, 1–13.
17. Singh, A. (2019). Environmental problems of salinization and poor drainage in irrigated areas: Management through the mathematical models. *Journal of Cleaner Production, 206,* 572–579.
18. Zhang, F., Guo, S., Zhang, C., & Guo, P. (2019). An interval multiobjective approach considering irrigation canal system conditions for managing irrigation water. *Journal of Cleaner Production, 211,* 293–302.
19. Harold Robinson, Y., & Golden Julie, E. (2019). MTPKM: Multipart trust based public key management technique to reduce security vulnerability in mobile ad-hoc networks. *Wireless Personal Communications, 109,* 739–760.
20. Nikzad, A., Chahartaghi, M., & Ahmadi, M. H. (2019). Technical, economic, and environmental modeling of solar water pump for irrigation of rice in Mazandaran province in Iran: A case study. *Journal of Cleaner Production, 239,* 118007.
21. García, A. M., García, I. F., Poyato, E. C., Barrios, P. M., & Díaz, J. R. (2018). Coupling irrigation scheduling with solar energy production in a smart irrigation management system. *Journal of Cleaner Production, 175,* 670–682.
22. Balaji, S., Harold Robinson, Y., & Golden Julie, E. (2019). GBMS: A new centralized graph based mirror system approach to prevent evaders for data handling with arithmetic coding in wireless sensor networks. *Ingénierie des Systèmes d'Information*, *24*(5), 481–490.
23. Philip, L., Cottrill, C., Farrington, J., Williams, F., & Ashmore, F. (2017). The digital divide: Patterns, policy and scenarios for connecting the 'final few' in rural communities across Great Britain. *Journal of Rural Studies, 54,* 386–398.
24. Rowshona, M. K., Dlamini, N. S., Mojid, M. A., Adib, M. N. M., Amin, M. S. M., & Lai, S. H. (2019). Modeling climate-smart decision support system (CSDSS) for analyzing water demand of a large-scale rice irrigation scheme. *Agricultural Water Management, 216*(2019), 138–152.
25. Harold Robinson, Y., Golden Julie, E., Balaji, S., & Ayyasamy A. (2017). Energy aware clustering scheme in wireless sensor network using neuro-fuzzy approach. *Wireless Personal Communications*, *95*(2), 703–721.
26. Imran, M. A., Alia, A., Ashfaq, M., Hassan, S., Culas, R., & Ma, C. (2019). Impact of climate smart agriculture (CSA) through sustainable irrigation management on resource use efficiency: A sustainable production alternative for cotton. *Land Use Policy*, *88*(2019), 104113
27. Angelopoulos, C. M., Filios, G., Nikoletseas, S., & Raptis, T. P. (2020). Keeping data at the edge of smart irrigation networks: A case study in strawberry greenhouses. *Computer Networks, 167*(2020), 107039.
28. Harold Robinson, Y., Rajaram, M. (2015). Energy-aware multipath routing scheme based on particle swarm optimization in mobile ad hoc networks. *The Scientific World Journal*, 1–9.

Blockchain-Based Cyberthreat Mitigation Systems for Smart Vehicles and Industrial Automation

Gunjan Madaan, Bharat Bhushan, and Raghvendra Kumar

Abstract The smart vehicle interaction grid is susceptible to cyberthreats, which are hard to crack employing outdated centralized security methodologies. Blockchain is an absolute P2P circulated record having cryptographically protected information. Blockchain demonstrates effective use cases in economic purposes, smart communication, etc. It expands to every industry comprising protected IoT devices. The exceptional characteristic of blockchain is that its distributed, unalterable, inspect database that safeguards transactions by keeping secrecy. In this paper, we anticipate the situation of the smart vehicle interaction grid and deliver considering approaches for assembling a blockchain-oriented system among smart vehicles. The paper discusses various use cases of blockchain in smart vehicles and ongoing advancement from motorized industries in addition to educational organizations. Blockchain in smart vehicles functions as a public distributed ledger which is capable of recording dealings between two groups. Though, the acceptance and execution of blockchain technologies in numerous organizations and facilities is a difficult job. The goal of this paper is to classify the obstacles for the effective execution of blockchain technologies in the companies.

Keywords Internet of things (IoT) · Industrial IoT · Blockchain · Distributed ledger · Security · Privacy · Smart vehicles · VANETS

G. Madaan
HMR Institute of Technology and Management (1) School of Engineering and Technology, Sharda University, India (2) and Department of Computer Science and Engineering, GIET University, Gunupur, India (3), New Delhi, India
e-mail: gunjanmadaan06@gmail.com

B. Bhushan (✉)
School of Engineering and Technology, Sharda University, Noida, India
e-mail: bharat_bhushan1989@yahoo.com

R. Kumar
Department of Computer Science and Engineering, GIET University, Gunupur, India
e-mail: raghvendraagrawal7@gmail.com

R. Kumar et al. (eds.), *Multimedia Technologies in the Internet of Things Environment*, Studies in Big Data 79, https://doi.org/10.1007/978-981-15-7965-3_2

1 Introduction

In the growing era of evolution, the technology used in automobiles is growing day by day with its advancement toward achieving the goal of autonomous vehicle for improving various factors such as accidents due to the human error, traffic flow efficiency, fuel economy, etc. and further approaching toward providing charging stations for upcoming electrical vehicles for reducing carbon emissions and stoppage of exploitation of fossil fuels [1, 2]. Intelligent vehicles use the features of Wireless internet which are combined with the different elements of communication, which make the vehicles a computer system themselves which functions similar to smartphone, controlled by the internal vehicular network and a wireless network of communication oriented complex computer [3]. For its rational functioning, prominent technologies are offered by this computerized structure such as self-driving features of an autonomous vehicle or advanced system assisted by the driver while also making it insecure and exposed to numerous types of cyberattacks and hacking [4, 5]. The intelligent vehicles are advancing swiftly from the outdated mechanical system which was used traditionally, to a cyber-physical system [6]. Wireless internet connectivity-based vehicular cloud systems urge dependence on the network software leading to major issues of safety and cybersecurity in the automotive sector. One of the crucial firms based on the above-mentioned facts is the use of a trusted network associated with the intelligent vehicle to secure the network from malicious cyberattacks. The vulnerability of the standard integrated security system is when it is compromised to the cyberattacks, frauds, etc., it affects the complete network system. While protecting the intelligent vehicle from the vulnerability of cyberattacks, the traditional privacy and security methods are more likely to be ineffective [7].

We usually characterize the linked vehicle which is supplied a connection of wireless network and likewise with vehicular ad hoc NETwork (VANET) [8]. Intelligent vehicles incorporate various control unit networks with sensors ensuring diverse vehicle utilities. The essential equipment recognizing smart and interconnected vehicle functionality can be characterized into working in or carried in an associated system. Interconnected cars have become an internet connected computer system like smartphones, while having very difficult control of moving forces that gives ease and suitability to occupiers.

At times, VANET is implied as the wireless system acknowledged as intelligent transportation system (ITS) [9, 10], which is planning to empower secure, well-organized stream of traffic, ecologically cognizant smart delivering facilities, along with additional traveler data facilities. ITS is intended to help the vehicle to distribute or to interchange data among vehicles and organizations through the network protocols, for example, vehicle-to-network (V2N) [11], vehicle-to-infrastructure (V2I) [12], vehicle-to-vehicle (V2V) [13] data exchange. C-V2X is a growing innovation created inside the 5G LTE cellular network by the normalization association of 3GPP (the third Generation Partnership Project) [14].

Self-driving vehicle [15] is the cutting-edge system where incitation and dynamic thinking work simultaneously while adapting to the approaching ecological data from. The beginning of a wireless network-based automobile cloud tends to initiate an amazing increment in the dependence on the connected software, subsequently, heading to concerns of cyberthreats for information safekeeping, compromising secrecy along with the purpose of keeping the vehicles safe. Main characteristics of existing and possible uses of the connected vehicle comprise of the following.

- Self-driving/driverless car: Evolved smart vehicles will be destined to help automatic self-driving without a person's contribution for its movement; these driverless automobiles present individuals a modern feel and assistance in both public and private transportation. These self-driven vehicles might be a harmful invention, disturbing existing organizations and redesigning metropolises, and giving new choices to do something else while driving a vehicle. A few instances of easily implied applications are automated valet facilities and remote charging facilities of electric cars.
- Automated cabs: Conventional traveling works in two different ways: Public transportation with a low repair fee at less ease and private transport at a significant high expense, however high comfort. With the approach of on-request automated traveling facilities, clients have profited by lower expenses and high comfort.
- Traffic efficiency: Enhancement of the course by considering the information about natural and traffic situations offers significant advantages in making the traffic efficient by improving traffic streams and eco-friendliness and decreasing traveling period.
- Traffic safety: Facts tell that over 90% of car crashes are often due to inaccurate judgment of drivers. The smart vehicle will noticeably get better at traffic well-being and successfully decrease the number of victims and injuries by applying safekeeping associated utilities that take advantage of surrounding data from in-vehicle sensors and mounted connection gadgetry.
- Mobile office: Smart vehicle exempts drivers from focusing on driving. Rather, they can relax or carry out tasks which are useful. Self-driving vehicles might at last transform into a fully functional moving office so clients can deal with easy or difficult things while traveling.

In summary, the major contribution of our paper is as follows:

- This paper presents the study of blockchain and introduces blockchain technology in industrial automation and smart vehicles.
- This paper discusses the issues of cyberthreats and how blockchain technology has resolved various threats and established security measures for smart vehicles.
- This paper redefines the issues occurring in distributed ledger technology (DLT) and concept of data privacy in blockchain.

The remainder of the paper is structured as follows. Section 2 explores various types of security threats in smart vehicles. Section 3 presents the overview, introduction of the automotive sector, smart vehicles and networking using blockchain technology. Sections 4 and 5 discusses how blockchain resolves the data privacy issues followed by numerous difficulties faced in establishing blockchain technology. Section 6 concludes the introduction of blockchain technology in automation. Figure 1 depicts the overall structure of the paper.

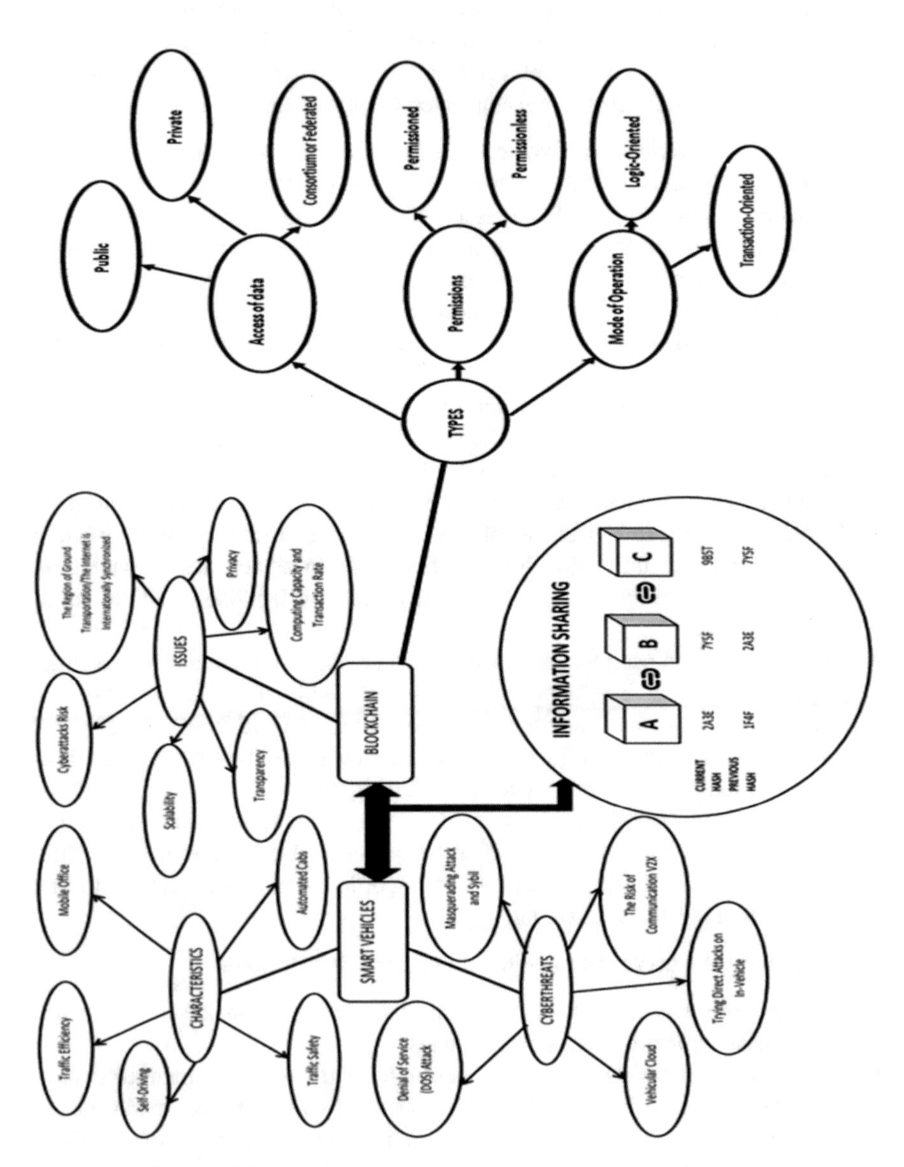

Fig. 1 Flow diagram of the overall paper

2 Cyberthreat Problems in Smart Vehicles

Smart vehicles are receiving a progressive development in academia and industry, but in order to make the most out of smart vehicles in relation with data innovation, objectifying a safe and secure way among these smart vehicles is needed. The wireless network finds regularity by making corresponding segments of communication the basic features of smart vehicles. Initially, the infrastructure of the small systems relies upon a central station that sorts out every communication aspect [16]. Second, uniquely selected systems ad hoc give related jobs [17, 18] and third, the initial part got organized with the hybrid network (hybrid VANET) [19]. For the communication between the infrastructure and vehicle, the ad hoc network is used [20]. They communicate with their advanced communication medium such as satellite and navigation, smartphones making them the most important component of IoT.

Linked smart vehicles offer a variety of refined facilities that act as an advantage to the vehicle drivers, clients, specialists of transport supervision, automanufacturers, and various facility contributors. At whatever point, a gadget which is associated with the internet is vulnerable to different cyberattacks threats. Future vehicles equipped with electronic tech have the most modernized characteristics and are connected to correspond with one another, which opens up new regions for data breach. Identical to the other ICT organizations that are frequently targeted for hacking, the smart vehicles must take into account the numerous potential outbreaks surface that programmers can use to disrupt into a system. On top of the networked links, other network components of the vehicle are exposed to security risks such as leakage of important private data, which may result in deadly outcomes. The outbreak surfaces of smart networked vehicles are wide-ranging and may expand significantly ahead of hacking the vehicle itself which is vulnerable and possible targets. Security threats of smart vehicles are as follows.

2.1 *Denial of Service (DoS) Attack*

Denial of service (DoS) attack tends to restrict clients from getting information or facilities in the computer systems. In vehicular systems, DoS attack makes use of numerous irrelevant messages to flood the traffic thereby adversely manipulate the network communications between nodes, onboard units, and side of the road units. Since the vehicle suffering an onslaught entering a smart city is a threat situation, there are a number of powerful processing databases close to the objective. An invader can probably utilize these for progressive attacks against the objective's stable hardware to counter the objective vehicle's capacity to recognize false messages by validating with its nearby data source.

2.2 Vehicular Cloud

Wireless networks change smart vehicles into a cyber-physical arrangement and every vehicle acts as a sort of node in the IoT. At that point, the vehicles are exposed to the possible malware attacks via different networks to get a gateway to the vehicle's sensitive controller or network bus of the internals. This possibly opens smart vehicles to a wide variety of safety, security, and confidential risks, for example, remote controlling of the vehicle, malicious attacks, unapproved connections, and location tracking. This cyberthreat is the largest hazard threat for smart self-efficient driving vehicles.

2.3 Masquerading Attack and Sybil

In a masquerading attack, a vehicle fakes its character and pretends to be real in the vehicle network. Outsiders can lead attacks, for example, sending false messages. In this, multiple locations are claimed by the vehicle due to which traffic chaos is created. In a sybil attack, the invaders produce various personalities and act as different genuine vehicles simultaneously [21]; this attack can compromise large use of the internet in vehicles. These lead to denial of service together.

2.4 The Risk of Communicating V2X

V2X interactions can enforce certain severe risks not in favor of vehicle safety and security and can result in upset traffic flow. Smart vehicles need to draw data, for example, the traffic condition, momentary traffic circumstances at uncalculated spaces, and information for collective driving. V2X interactions are open to invasions by transferring false data or altering messages. Another difficulty with blockchain innovation for smart vehicles is the establishment of a safe V2X transmission system to believe messages from different vehicles or surroundings, along with not giving away on security.

2.5 Wormhole Attack

Every network or system which is highly dependent on routing algorithms and chooses an efficient path for data flow in a network is highly vulnerable and can lead to system failure when attacked by a wormhole. This attack has two or more malicious nodes as targets to cause congestion or jamming in the system data flow by hiding the true path length between these nodes and fake as the shortest and efficient routing path for data flow [22, 23].

2.6 Replay Attack

Replay attack creates congestion in the system with its unique feature of replaying large amounts of messages simultaneously thereby, jamming the network and disrupting the priority messages, leading to dropping off from the queue. This attack considerably decreases the system efficiency and the usage of bandwidth therefore, increasing its cost. Digital signature technology like message forgery is also unable to stop or prevent this system activity.

2.7 Malicious Software (Malware)

Malware attacks like viruses or worms come from unauthorized sources or software and firmware updates. They can also affect the vehicle system by taking control of the main system frame and can disrupt or disable its essential services. This malware can also come from the vulnerable computers of servicing stations which maintain and diagnose vehicles during servicing. Therefore, malware spreads from one system to another via a link of software while updating or servicing it from unauthorized or low-security systems.

Table 1 presents a summary of various aforementioned attacks.

3 Blockchain Overview

Blockchain has been established as a developing innovation with the potential to disturb every single conventional industry. The effect of blockchain innovation has been felt in the business segment as of now with digital forms of money like cryptocurrencies such as Ethereum and bitcoins. Right now, the blockchain innovation has been taken advantage of in applications, for example, social services, therapeutic and smart contacts, and business purposes; afterwards, efforts have been made to integrate blockchain into much complicated regions, for example, the security and privacy of IoT.

Table 1 Summary of attacks in smart vehicles

S. No.	Types of attacks	Description
1	Denial of service (DoS)	It restricts the smart vehicles stable hardware for recognizing the false message by sending a large number of irrelevant messages and exceeds the capacity to process them. Therefore, it can lead to disabling or malfunctioning of vehicle systems
2	Vehicular cloud	Smart vehicles consist of wireless networks which make them vulnerable to cyberthreats and an invader can access the network via multiple gateways and can lead to various security risks
3	Direct attacks on in-vehicle	It commands the core utilities of a vehicle via entering the in-vehicle system by creating an unapproved network to make connections and to gain access to its system
4	Masquerading and sybil	Masquerading attack fakes vehicle characteristics and pretends to be the original vehicle in the network so that it gets easy for an invader to attack the network. Sybil attack simultaneously creates multiple personalities and pretends as different genuine vehicles
5	The risk of communicating V2X	V2X interaction can enforce various safety and security risks and is open to invasions by transferring false data or altering messages
6	Wormhole attack	It hides the true path length between two or more malicious nodes therefore, disrupting the routing algorithm with wrong information and causing congestion in the system
7	Replay attack	It runs a system activity of replaying and deleting large amounts of messages therefore, decreasing system efficiency, increasing cost of bandwidth and leading to dropping off priority messages from the queue
8	Malicious software (malware)	It spreads from one system to another via a link of software or networks and attack system main frame to gain its control and disrupts system behavior by disabling essential services

3.1 Background

Blockchain innovation works for the digital currency, which has lately been utilized to develop trust and dependability in P2P systems with related topologies as systems of smart vehicles. Blockchain has demonstrated effective use cases in money related applications, technical contact, and distributing media content with advanced patent security [24]. An unrivaled component of blockchain is its distributed, unchallenge-able, auditable record for safe transfers with confidentiality reserved [25]. Right now, we consider the reliable background of smart vehicle data transmission systems and how blockchain innovation generates trust among smart vehicles.

3.2 Types of Blockchains

Depending on user interactions and management of data, different types of blockchains are used in industrial applications. Different blockchains that exist are detailed in subsections below.

3.2.1 Public Blockchains

Applications where high transparency and user interaction is required without any restrictions on publishing or validating transactions by users, public blockchain is useful in industrial applications. Public blockchains are completely decentralized and permission less as users do not require any access permission from a source, and therefore, all users can perform actions and can maintain a copy of the blockchain. Each transaction involves some fees which act as an incentive for others who are attempting to create new blocks in the blockchain. This gives an advantage over security as it will be costly to hack the blockchain and tamper its content.

3.2.2 Private Blockchains

A single owner operates and maintains the blockchain and grants users' access to the network and decides which nodes will be validators. Due to single owner, private blockchain might not be considered as decentralized blockchain. Private blockchains are permissioned as an owner provides permission to access it and is suitable for a single enterprise. There is no transaction fee attached, as the blocks are created or published by a node within the network, so private blockchains have low tamper resistance as compared to a public blockchain.

3.2.3 Consortium or Federated Blockchains

Consortium blockchain is permissioned and partially decentralized which is a bit similar to a private blockchain where the user is granted access to the network by the owner. But in consortium blockchain, there is a group of owners/consortium members rather than a single owner which keeps an account of the data exchanged between the parties in a synchronous manner. Nodes are pre-selected for validation

Table 2 Comparison between blockchain infrastructure

S. No.	Properties	Public	Private	Consortium
1	Access	No restriction on accessing the blockchain	Restricted	Restricted
2	User Identity	Unknown/anonymous	Known participants of organization	Known participants of multiple organizations
3	Participation type	Completely Public/all nodes	Only members/nodes of single organization	Only pre-selected members/nodes of multiple organization
4	Type of permission	Permission less	Permissioned	Permissioned
5	Tamper resistance	High	Low	Low
6	Transaction speed	Slow, due to large amount of public access	Fast, as access is restricted to single organization	Fast, as access is restricted to pre-selected members of multiple organization
7	Decentralization	Completely decentralized	Not decentralized	Partially decentralized

operation and perform validation within the network, so there is no transaction fee involved and publishing a new block is also reasonable.

Table 2 presents the comparison of these aforementioned types of blockchain technology.

3.3 Inherent Characteristics of Blockchain

Every network or system requires some technology which inherits several characteristics to fulfill their requirements so that the user can trust the network and can process various transactions on it. Blockchain technology consists of various characteristics which involve transparency, finality, consensus, etc., making it suitable for different types of networks and is proved to be a trustworthy technology. The characteristics of blockchain making it suitable for smart vehicles are depicted in Fig. 2 and listed as follows.

- Provenance: Each member is aware of the exact location from where the merit originated from and in what way its possession has changed during exchanges.
- Immutability: No member can alter or exchange once ceit has been logged in the record (Ledgers). Even if an error occurs in ceit, it cannot be removed or altered

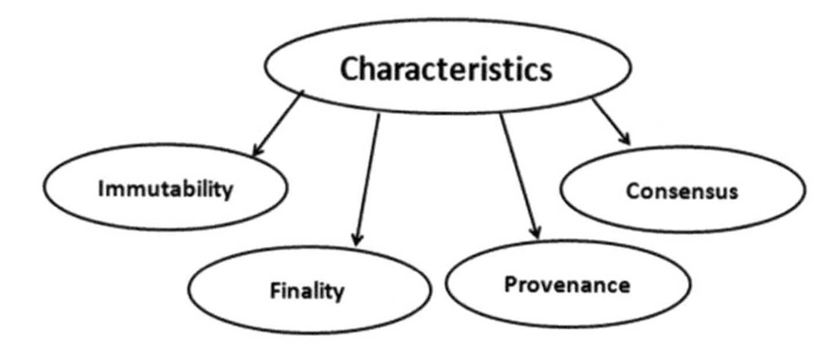

Fig. 2 Characteristics of smart vehicles based on blockchain

as it has been already recorded, only a new ceit can take place but still the previous error will be visible in the records.
- Finality: There exists only one ledger in the network where all the records of each transaction along with every details of the transaction are stored and cannot be altered. Therefore, ledger shows the character of finality in the blockchain network.
- Consensus: Every participant must agree to follow a set of rules then only their transaction will be accepted and recorded on the blockchain. This is known as consensus. Due to this characteristic of blockchain, every participant trusts this network and knows if the transaction violates the defined rules of the network, it will be considered as invalid transaction.

3.4 *How Blockchain Works on Trusted Networks*

A blockchain is a distributed record that consists of information about all exchanges performed over a P2P system. It comprises a lot of blocks that store specific data and are connected with cryptographic protocols. A huge factor of blockchain is that once a client records the information inside it, making changes later on turns out to be extremely troublesome. Blockchain utilizes an eminent irregular public/private key and hash cryptography system to authorize the validation of exchanges. The personal key is utilized to authorize the exchanges and every mining node gathers new exchanges and solution to the agreement. At the point, when a node obtains an agreement of the given system, it communicates the details to all nodes. The principal node that comprises the details effectively, i.e., finds the agreement's solution, acquires a prize of digital currency which turns into the key motivation for members to keep up to the blockchain without failure. When it comes to a hash cryptography system to authorize the validation of exchanges, at the point when someone looks into a solitary block, it consists of the application information, the hash estimation of the past block, and hash of its own. Hashes are at all times distinct and assist to recognize a block and its constituents. The hash number is made with the assistance of a hashing equation that obtains a random number of I/P and produces a definite

O/P. Frequently, the hash number of a following block is created utilizing the hash of the preceding block along with a bit of its own exchange.

4 Blockchain to the Rescue

Blockchain can ensure three significant threats to form a reliable system among smart vehicles with no significant control organization, ensuring the secrecy and protection from cyberattacks, and assuring flexibility under vulnerabilities or adverse repercussions on the vehicular cloud system [26, 27]. Blockchain innovation depends upon P2P networks that permit smart vehicles to connect with one another with no interference of third reliable agencies. Regardless of whether a few nodes are disconnected or compromised, blockchain be capable of making the vehicular system work normally. All exchanges in blockchains are time-marked and encoded with private keys, so without much effort, vehicle clients can track the previous exchanges and trace accounts at whichever period of time with ease. Additionally, this characteristic lets every member to track and follow the source of information or resources, that is, the authorized data concerning the provenance of the resource and in what way its possession has reformed through exchanges. This following ability avoids possible attack by transferring fake information or altered messages in V2X networks, protecting the system against fake security. An invader may try to deanonymize the identification of the automobile or tracing confidential data of the client, and every exchange comprises highly classified information that threatens the security of the client. The blockchain can guarantee safety regarding secrecy of the client by means of a hash equation and encoding utilizing unbalanced cryptography [28, 29].

4.1 Blockchain in Vehicular Networks

We start by taking into account the safety and secrecy necessities for vehicular ad hoc systems. In rising vehicular ad hoc systems for self-driving and smart vehicles, there is a necessity for vehicles to interact with each other, and using roadside components which give interactive assistance, to get data and updates about the status of their environment (e.g., street works), and to convey their activities. Safety characteristics for ITS secure messaging occur from two primary significant prerequisites—privacy and safety—which are clashing here in a number of aspects: for safety uses, striking indications, for instance, we need that messages are transmitted much often, that their texts are exact, unchanged, and faithful, and that they are collected by each one of the individuals who need the data they include. We, in this manner, need that a message's dispatcher is, and its message is, trusted by its receivers. Secrecy necessities, nevertheless, drive a necessity for identification of message dispatchers' actual characteristics cannot be obtained from the security messages. The chance to decentralize entry into exchanged data over these systems is very much beneficial.

This could stop any single group from accessing all data and hence tracking the activities of groups and could imply that no single group is in charge for the saving and managing data, indicating an only place of breakdown [30].

4.2 Blockchain in Automotive Sector

Vehicle organizations have continuously acquired revolutionary advanced answers to crack difficulties in areas, for example, production, coordination, deals, supply chain, client services, and all regarding it. With 4.8% yearly development, this business witnesses fast grip across topographies and is a hotbed for technology advancements. Blockchain can profit the vehicle business from various methods. Utilizing smart contracts and IOT, vendors can computerize numerous procedures engaged with vehicle deals, administration, guarantee request processing, and many more. Also, it helps the purchaser and vender to evade mediators or outsiders to engage in exchanges or protection claims. This decentralized record innovation can assist in building a condition of calmness in clients, makers, and maintenance stations with its clarity and rigid character. The utilization of blockchain in the automotive business goes from cryptotokens to give loyalty prizes to their clients that can hold the reliability or source of the vehicle. Keeping up an unalterable log of all records, for example, administration logs, information about the damage of extra parts and so on can assist the maker with the estimation of the real resale cost of the vehicle. The following section explore blockchains affecting various sections of automotive sector.

4.2.1 Blockchain in Vehicle Production

- Supply chain: After the production of the vehicle, its parts could be damaged, lost, or stolen while delivering or may be replaced. Blockchain comes to the rescue and helps to track down those parts and could be able to control if any modifications like replacement, etc. are made or even if a defect occurs at the time of manufacturing of parts.
- Production process: Blockchain technology also benefits in keeping track of various processes during the production of the vehicle. It is used to keep a record of billing data, multiple quality inspections along with the assembly information at each step till the production is completed.
- Smart insurance: Using blockchain technology, all the data from sensors installed in vehicles are documented and stored in blockchain "black box," which can help in resolving accident cases as the data or document stored in the blockchain cannot be manipulated. It can also help in tracking rightful owners of the vehicle and can also help in claiming insurance in case of vehicle sharing.

4.2.2 Blockchain for Vehicle Owner

- Buying or selling a vehicle: With the help of blockchain, there is major transparency while buying a vehicle, for example, vehicles history can be easily verified and can take an overview of vehicle parts and need for repairs. Due to blockchain-based smart contracts, there is no middleman; therefore, sellers, and buyers can make a deal and transaction directly.

4.2.3 Blockchain in the Mobility Sector

- Leasing and Vehicle Financing: Blockchain has made numerous processes of leasing and financing of vehicle sections safe and optimized with the help of smart contracts. The vehicle system can be deactivated thereby; users cannot access its vehicle if the lease rate of the financed vehicle is not paid on time.
- Carsharing: Blockchain helps in storing the identity of owners along with their vehicle settings safely. Blockchain can also enable car sharing users to be registered once rather than registering owners separately for the same vehicle.
- Car rental company: Blockchain enables monitoring of vehicle maintenance for rental companies to help them follow up with various vehicles simultaneously.

4.3 Smart Vehicles Based on Blockchain

In today's secure data exchange, all the information exchanged is encoded and verified utilizing a shared/personal key infrastructure with permit-based joint verification between cloud servers or connected nodes. Though joint verification guarantees that the two contraptions can confirm the validity of one another, these conventional security approaches are yet not liable for safe information sharing among smart vehicles [31]. A major necessity of ITS is the confirmation of immutability, provenance, and an error-resilient system between smart vehicles.

4.4 Blockchain for Industry 4.0 Applications

Despite being a very useful tool for Industry 4.0 Applications, blockchain is not essential to be installed for every problem. There are many technologies which serve the same purpose, for example, high-speed transactions, infinite scalability or to create trustless decentralized networks, etc. Example of such technologies is tangle which implements a directed acyclic graph (DAG), which is used to represent all kinds of information in a structured way. Along with the multiple features, tangle has some drawbacks also like it is not much tested or verified due to less usage by users or developers, low-security issues or uses a synchronization node which affects decentralization. Therefore, there are some factors or frameworks which can help to decide whether to use blockchain or other alternative technology. They are as follows.

- It is beneficial to install blockchain technology where a decentralized network is required by the application.
- In industry application, large amount of data is shared or exchanged among multiple users. So, if P2P communication is required, blockchain will benefit the purpose.
- Blockchain is useful where robust infrastructure, privacy, or security of information is essential in an industry application.

5 Issues in Distributed Ledger Technology (DLT)/Blockchain

5.1 Scalability

Scalability implies the ability of the blockchain system to deal with and supervise an expanding quantity of DLT. Balancing the blockchain is a real test and has been a dynamic zone of study for quite a while. Scalability is the fundamental of blockchain design for smart vehicles due to the fact that every vehicle has a restricted quantity of loading capacity and energy funds to work for a mining node or circulated record [32]. In the conventional blockchain system, every node utilizes huge amounts of energy for mining, and an information stock server is essential to store collective information of the circulated record permanently, which is practically impossible to execute in a vehicle ecosystem. Utilization of DLT depends severely on the assigned block dimension of the transferred data, velocity of transferring data through the system, the hidden proof-of-work procedure and the confirmation of miner data on each node. The bigger block dimension will instantly let a larger amount of exchanges and encourage a scalable distributed ledger technology usage in different organizations and facilities other than cryptographic forms of money [33].

5.2 Computing Capacity and Transaction Rate

To include another exchange, consent of group members must be fed into the circulated record of the blockchain. In any case, the executed blockchain consent method still does not offer noticeable enhancements to exchange velocities. In the situation of frequently utilized procedures, the time of consent to evaluate a new block is ten minutes for confirmation of work (for bitcoin) and seventeen seconds for verification of stake (for Ethereum).On the contrary, the normal approval time of bitcoin is around two hundred fifty minutes; the highest time noted at any point was over twenty-five hundred minutes (roughly forty-two hours) in February 2018. These prevailing consent procedures of the blockchain are very time-consuming for smart vehicles to take messages among vehicles instantaneous. In addition, mining nodes utilize a significant amount of processing energy. To solve the obstacle of velocity and energy resources, certain new, creative methodologies must be presented to enhance these fundamental blockchain properties.

5.3 Transparency

In spite of incredible characteristics of the blockchain innovation, numerous modifiers are being opposed by the sort of information that might be there inside the circulated record. Numerous individuals believe that encoded data handled with shared and personal keys are supposed to back up the classified data being recorded on the blockchain. In the decentralized system, all groups circulate data. No confidential information is supposed to be put on a circulated record, encoded, or something else. Distributed records are supposed to include the absolute minimum data, understood just by those with the necessity and approval to know, to allow announcement, coordination, and authorization. For privacy, secret and delicate information is supposed to exist on a blockchain as it offers an additional level of safety with unchangeable records.

5.4 Privacy

In a P2P distributed ledger technology network, there is a major concern for the operator in a blockchain exchange. In personal blockchain, the work is done on the mutual trust as the identity of the person is not disclosed to each other. Instead of being public for clear business methods, it works as a private association. From a safety view point, scientists created different methods focusing on secrecy concerns concentrated on private information. Information anonymization techniques try to secure privately recognizable data [34]. We consider two sorts of privacy, namely privacy of information and privacy of the party. If a party A presents a message to the blockchain, A may not desire the other parties of the system to realize that he has posted right now, or capable of connecting this post to different posts by A.

5.5 The Region of Ground transportation/The Internet Is Internationally Synchronized

Continuous combination of neighboring vehicular systems and worldwide dispersed collection platforms represent various difficulties since all these construction blocks are dissimilar as far as they rely on the foundation and programming components. Land transportation is restricted to local areas, yet the blockchain is worldwide cyberspace. Nearby exchanges are not required to be disclosed on the internationally circulated record, which may decrease unwarranted information traffic. The issue of the neighborhood and internationally circulated record can be resolved by concentrating on information-driven features of vehicular systems. The discovery of the irregularity of cyberattacks for different areas by proof of driving is a possible remedy.

5.6 Cyberattacks Risk

Attacks which are network oriented such as a DoS attack might be able to establish interruptions and jamming in the blockchain peer-to-peer network [35]. A DoS attack intends to disturb the steady procedure of the blockchain system by filling the nodes with false appeals. Since the vehicle undergoing an attack is a case of the much broader base of the smart city condition, there are many powerful offices nearby to the objective. A hacker might probably utilize these for doing attacks against the objective's established solid components to counteract the objective's vehicle's capacity to recognize fake messages through authentication with its nearby data sources.

6 Future Issues and Challenges

The administrative trade-offs, challenges, and future research scope toward blockchain integration in industrial IoT and smart vehicles are listed in the following section.

- Blockchain technology requires high computational power and capacity for fast processing without decreasing the security. Client/server system is to be replaced by blockchain technology, but the data is stored in nodes or the IoT devices with low computational and very low storage capacity. Therefore, it can become a barrier in the establishment of this technology.
- There is a need to upgrade IoT devices to make them compatible with upcoming high-speed networks and systems.

- In blockchain technology, the block mining is concentrated on large computations; therefore, prerequisite energy and processing time of blockchain-bases IoT devices need to be analyzed.
- There is a need to develop a dynamic security framework which is versatile for both high-end systems and low powered devices like IoT devices.
- Functioning of blockchain decreases as there is an increase in the statistics of nodes which are not addressed.
- This technology requires proper standardized policies to achieve complete interoperability without compromising trust and security measures.

7 Conclusion

In this paper, we provide insights to the bibliophiles regarding the enduring growth of intelligent vehicles and their security against cyberthreats, the concept of blockchain-based VANET, intelligent transportation systems (ITS), and IoT networks. It is very tricky to utilize this technology in vehicle cloud systems as it may be inefficient to imitate vehicle networks as blockchains and enhance them; demands the complete replacement of prevailing systems. Blockchain certifies protection and flexibility under repercussions or attack on the vehicular cloud control. All exchanges in blockchains are time-stepped and scrambled with private keys. It can guarantee the assurance of the protection of the client by utilizing a hash capacity, and encryption utilizing uneven cryptography. The chance to decentralize access to communications data over these systems could keep any single gathering from survey all data and along these lines following the activities of gatherings. Yet there are many challenges such as organizational issues, economic issues, environmental issues, and other social factors that lead to further adoption of blockchain technology and finally positioning DLTs.

References

1. Hasan, M. G. M. M., Datta, A., & Rahman, M. A. (2018). Poster abstract: Chained of things: A secure and dependable design of autonomous vehicle services. In *2018 IEEE/ACM Third International Conference on Internet-of-Things Design and Implementation (IoTDI)*. https://doi.org/10.1109/iotdi.2018.00048.
2. Zhao, N., & Wu, H. (2019). Blockchain combined with smart contract to keep safety energy trading for autonomous vehicles. In *2019 IEEE 89th Vehicular Technology Conference (VTC2019-Spring)*. https://doi.org/10.1109/vtcspring.2019.8746337.
3. Biswas, B., & Gupta, R. (2019). Analysis of barriers to implement blockchain in industry and service sectors. *Computers & Industrial Engineering, 136,* 225–241. https://doi.org/10.1016/j.cie.2019.07.005.
4. Axon, L., Goldsmith, M., & Creese, S. (2018). Privacy requirements in cybersecurity applications of blockchain. *Advances in Computers Blockchain Technology: Platforms, Tools and Use Cases*, pp. 229–278. https://doi.org/10.1016/bs.adcom.2018.03.004.

5. Jin, P. J., Zhang, G., Walton, C. M., Jiang, X., & Singh, A. (2013). Analyzing the impact of false-accident cyber attacks on traffic flow stability in connected vehicle environment. In *2013 International Conference on Connected Vehicles and Expo (ICCVE)*. https://doi.org/10.1109/iccve.2013.6799866.
6. Nawa, K., Chandrasiri, N. P., Yanagihara, T., Komori, T., & Oguchi, K. (2012). Cyber physical system for vehicle application. In *2012 IEEE International Conference on Cyber Technology in Automation, Control, and Intelligent Systems (CYBER)*. https://doi.org/10.1109/cyber.2012.6392540.
7. Singh, A., & Singh, M. (2018). An empirical study on automotive cyber attacks. In *2018 IEEE 4th World Forum on Internet of Things (WF-IoT)*. https://doi.org/10.1109/wf-iot.2018.8355124.
8. Kaur, R., Singh, T. P., & Khajuria, V. (2018). Security issues in vehicular ad-hoc network (VANET). In *2018 2nd International Conference on Trends in Electronics and Informatics (ICOEI)*. https://doi.org/10.1109/icoei.2018.8553852.
9. Chaqfeh, M., Mohamed, N., Jawhar, I., & Wu, J. (2016). Vehicular cloud data collection for intelligent transportation systems. In *2016 3rd Smart Cloud Networks & Systems (SCNS)*. https://doi.org/10.1109/scns.2016.7870555.
10. Ashtankar, P. P., & Dorle, S. S. (2015). Application based design strategies and simulation of wireless adhoc communication network using intelligent transportation system. In *2015 International Conference on Energy Systems and Applications*. https://doi.org/10.1109/icesa.2015.7503450.
11. Kim, S. (2018). Blockchain for a trust network among intelligent vehicles. *Advances in Computers Blockchain Technology: Platforms, Tools and Use Cases*, pp. 43–68.https://doi.org/10.1016/bs.adcom.2018.03.010.
12. Miller, J. (2008). Vehicle-to-vehicle-to-infrastructure (V2V2I) intelligent transportation system architecture. In *2008 IEEE Intelligent Vehicles Symposium*. https://doi.org/10.1109/ivs.2008.4621301.
13. Mallikarjuna, G. C. P., Hajare, R., Mala, C. S., Rakshith, K. R., Nadig, A. R., & Prtathana, P. (2017). Design and implementation of real time wireless system for vehicle safety and vehicle to vehicle communication. In *2017 International Conference on Electrical, Electronics, Communication, Computer, and Optimization Techniques (ICEECCOT)*. https://doi.org/10.1109/iceeccot.2017.8284527.
14. Maglaras, L., Al-Bayatti, A., He, Y., Wagner, I., & Janicke, H. (2016). Social internet of vehicles for smart cities. *Journal of Sensor and Actuator Networks*, 5(1), 3https://doi.org/10.3390/jsan5010003.
15. Duong, M.-T., Do, T.-D., & Le, M.-H. (2018). Navigating self-driving vehicles using convolutional neural network. In *2018 4th International Conference on Green Technology and Sustainable Development (GTSD)*. https://doi.org/10.1109/gtsd.2018.8595533.
16. Soni, S., & Bhushan, B. (2019). A comprehensive survey on blockchain: Working, security analysis, privacy threats and potential applications. In: *2019 2nd International Conference on Intelligent Computing, Instrumentation and Control Technologies (ICICICT)*. https://doi.org/10.1109/icicict46008.2019.8993210.
17. Perkins, C. E. (2008) Ad hoc networking; addison-wesley professional. Boston, MA, USA: Addison-Wesley.
18. Caballero-Gil, P., Caballero-Gil, C., & Molina-Gil, J. (2013). How to build vehicular ad-hoc networks on smartphones. *Journal of Systems Architecture, 59*(10), 996–1004. https://doi.org/10.1016/j.sysarc.2013.08.015.
19. Wang, M., Shan, H., Lu, R., Zhang, R., Shen, X., & Bai, F. (2015). Real-Time Path planning based on hybrid-VANET-enhanced transportation system. *IEEE Transactions on Vehicular Technology, 64*(5), 1664–1678. https://doi.org/10.1109/tvt.2014.2335201.
20. Tornell, S. M., Patra, S., Calafate, C. T., Cano, J.-C., & Manzoni, P. (2015). GRCBox: Extending smartphone connectivity in vehicular networks. *International Journal of Distributed Sensor Networks, 11*(3), 478064. https://doi.org/10.1155/2015/478064.

21. Maglaras, L. A., Basaras, P., & Katsaros, D. (2013). Exploiting vehicular communications for reducing CO2 emissions in urban environments. In *2013 International Conference on Connected Vehicles and Expo (ICCVE)*. https://doi.org/10.1109/iccve.2013.6799765.
22. Yan, G., Wen, D., Olariu, S., & Weigle, M. C. (2013). Security challenges in vehicular cloud computing. *IEEE Transactions on Intelligent Transportation Systems, 14*(1), 284–294. https://doi.org/10.1109/tits.2012.2211870.
23. Ji, S., Chen, T., & Zhong, S. (2015). Wormhole attack detection algorithms in wireless network coding systems. *IEEE Transactions on Mobile Computing, 14*(3), 660–674. https://doi.org/10.1109/tmc.2014.2324572.
24. Malik, A., Gautam, S., Abidin, S., & Bhushan, B. (2019). Blockchain technology-future of IoT: Including structure, limitations and various possible attacks. In *2019 2nd International Conference on Intelligent Computing, Instrumentation and Control Technologies (ICICICT)*. https://doi.org/10.1109/icicict46008.2019.8993144.
25. Arora, D., Gautham, S., Gupta, H., & Bhushan, B. (2019). Blockchain-based security solutions to preserve data privacy and integrity. In *2019 International Conference on Computing, Communication, and Intelligent Systems (ICCCIS)*. https://doi.org/10.1109/icccis48478.2019.8974503.
26. Singh, M., Kim, S. (2017). *Safety Requirement Specifications for Connected Vehicles*. https://www.arXiv:1707.08715.com.
27. Bhushan, B., & Sahoo, G. (2020). Requirements, protocols, and security challenges in wireless sensor networks: An industrial perspective. *Handbook of Computer Networks and Cyber Security*, pp. 683–713.https://doi.org/10.1007/978-3-030-22277-2_27.
28. Polyzos, G. C., & Fotiou, N. (2017). Blockchain-assisted information distribution for the internet of things. In *2017 IEEE International Conference on Information Reuse and Integration (IRI)*. https://doi.org/10.1109/iri.2017.83.
29. Sharma, T., Satija, S., & Bhushan, B. (2019). Unifying blockchain and IoT: Security requirements, challenges, applications and future trends. In *2019 International Conference on Computing, Communication, and Intelligent Systems (ICCCIS)*. https://doi.org/10.1109/icccis48478.2019.8974552.
30. Saini, H., Bhushan, B., Arora, A., & Kaur, A. (2019). Security vulnerabilities in Information communication technology: Blockchain to the rescue (A survey on Blockchain Technology). In *2019 2nd International Conference on Intelligent Computing, Instrumentation and Control Technologies (ICICICT)*. https://doi.org/10.1109/icicict46008.2019.8993229.
31. Sharma, P. K., Moon, S. Y., & Park, J. H. (2017). Block-VN: A distributed blockchain based vehicular network architecture in smart city. *Journal Information Processing System, 13*(1), 184–195.
32. Beer, C., & Weber, B. (2014). Bitcoin—The promise and limits of private innovation inmonetary and payment systems. *MONETARY Policy & the Economy, 4,* 53–66.
33. Zyskind, G., Nathan, O., & Sandy P. A (2015). Decentralizing privacy: Using blockchain to protect personal data. In *2015 IEEE Security and Privacy Workshops*. https://doi.org/10.1109/spw.2015.27.
34. Gervais, A., Ritzdorf, H., Karame, G. O., & Capkun, S. (2015). Tampering with the delivery of blocks and transactions in bitcoin. In *Proceedings of the 22nd ACM SIGSAC Conference on Computer and Communications Security—CCS15*. https://doi.org/10.1145/2810103.2813655.
35. Karame, G. (2016). On the security and scalability of bitcoins blockchain. In *Proceedings of the 2016 ACM SIGSAC Conference on Computer and Communications Security—CCS16*. https://doi.org/10.1145/2976749.2976756.

IT Convergence-Related Security Challenges for Internet of Things and Big Data

Y. Reeginal, G. Vijayarani, C. Anuradha, and Y. Harold Robinson

Abstract The smart device is connected with the Internet of things, and it shows the monitoring of several sensors with transmission parameters to produce the efficient system. The Internet of things has generated the data even though the transmission range is out of the range in spite of reduced cost and period. This chapter demonstrates the Internet of things- and big data-related applications and analyzes the various security issues whenever the environment is diagnosed. The latest security issues may happen in every element in the Internet of things because an element is communicated to the security vulnerability. Whenever the data is generated through the big data and the Internet of things surroundings, these kinds of problems are identified according to the security-related relationship. The necessity of the security parameters of the ICT-based environment is analyzed in detailed way.

Keywords Security issues · Internet of Things · Big data · IT convergence · Software · Automation system

Y. Reeginal
Department of Physics, SCAD College of Engineering and Technology, Cheranmahadevi, India
e-mail: reeginal1986@gmail.com

G. Vijayarani · C. Anuradha
Department of Computer Science and Engineering, SCAD College of Engineering and Technology, Cheranmahadevi, India
e-mail: vijirajaka@gmail.com

C. Anuradha
e-mail: anurajahpaul@gmail.com

Y. H. Robinson (✉)
School of Information Technology and Engineering, Vellore Institute of Technology, Vellore, India
e-mail: yhrobinphd@gmail.com

R. Kumar et al. (eds.), *Multimedia Technologies in the Internet of Things Environment*, Studies in Big Data 79, https://doi.org/10.1007/978-981-15-7965-3_3

1 Introduction

The Internet of things is the network of sensor devices such as sensor electronics, electronic appliances, and others embedded with electronics, software, sensors, and network connectivity that enables these devices to collect and exchange data [1]. The Internet of things network consists of sensors for sensing a particular situation or environment (sensor node), a processor for processing the collected information, and a data transmitting and receiving device (sink node) [2]. In Internet of things, a seamless communication and information delivery is achieved from a physical component such as a sensor to user service, so security vulnerabilities specific to each individual component may exist [3]. New security vulnerability may occur in each component in Internet of things, since a component is connected to security vulnerability [4]. Big data is difficult to manage and analyze by conventional methods, since its data format is diverse and unstructured and its distribution speed is fast [5]. The reason for doing research on security issues of both Internet of things and big data is that it may be more effective in achieving security measures when the relationship from the attributes of data generated in Internet of things and to that in big data is diagnosed [6]. This study derives and diagnoses security issues of both Internet of things and big data in the IT convergence environment [7].

The components of Internet of things as shown in Fig. 1 are humans, things, the Internet, and distributed services. Uniquely identifiable tools and devices are the range of things [8]. The range of things is uniquely identifiable tools and devices. The interlocking process of Internet of things is connected to sensing–networking–information processing–intelligent relationship–information exchange between things [9]. In order to apply Internet of things in real life, generic technologies need to be systematically implemented [10].

Generic technologies can be divided into the sensor and network hardware technology such as controllers and communication chips, the middleware software technology [11] for storing and analyzing the data received from things, and the application software technology that interprets, expresses, and processes data as meaningful results [12]. The Internet of everything has the revolution that connected with the people and Internet of things, and this process is called as the people to machine for producing the relationship [13]. The people and mobile have the relationship called as the people to people, and finally the machine-to-machine relationship is connected with the data and the Internet of things [14]. The entire process is demonstrated in Fig. 2.

The sensor network technology is the very basic technology that recognizes, extracts data from things, and transmits them to the Internet [15]. The role of human in Internet of things environment is to implement Internet of things and utilize the final information [16]. Human does not need to be involved on the operational stage, since the operation of things is automated [17]. The healthcare system that connects with the Internet of things through the smart devices like gateway device and smartphone with sensors is placed to observe the condition [18]. The user's health is monitored using the smartphone and all the details are updated through the network connectivity,

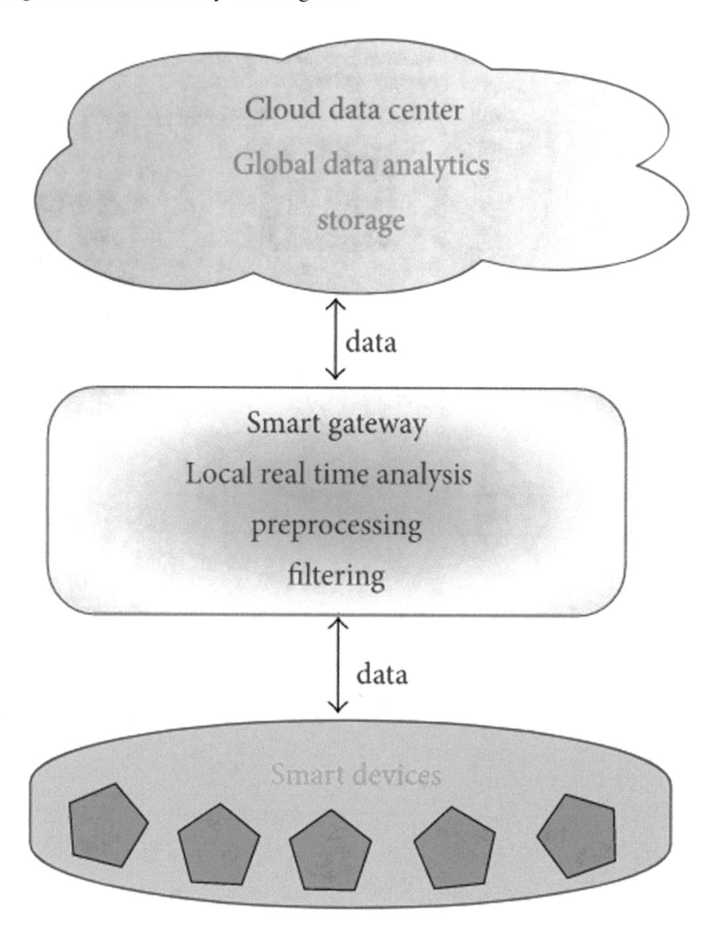

Fig. 1 Components of Internet of Things

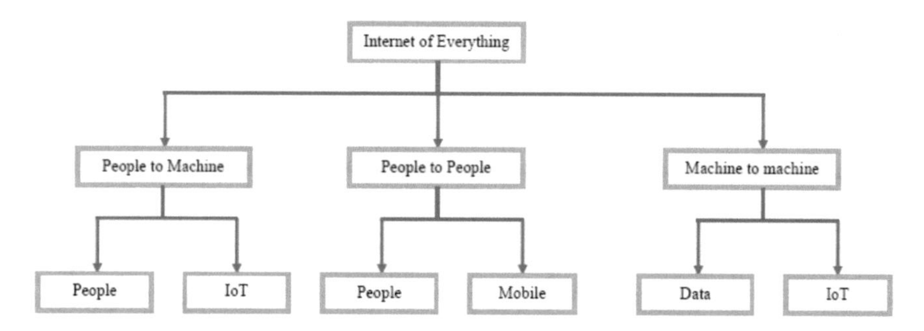

Fig. 2 Revolution of Internet of Things

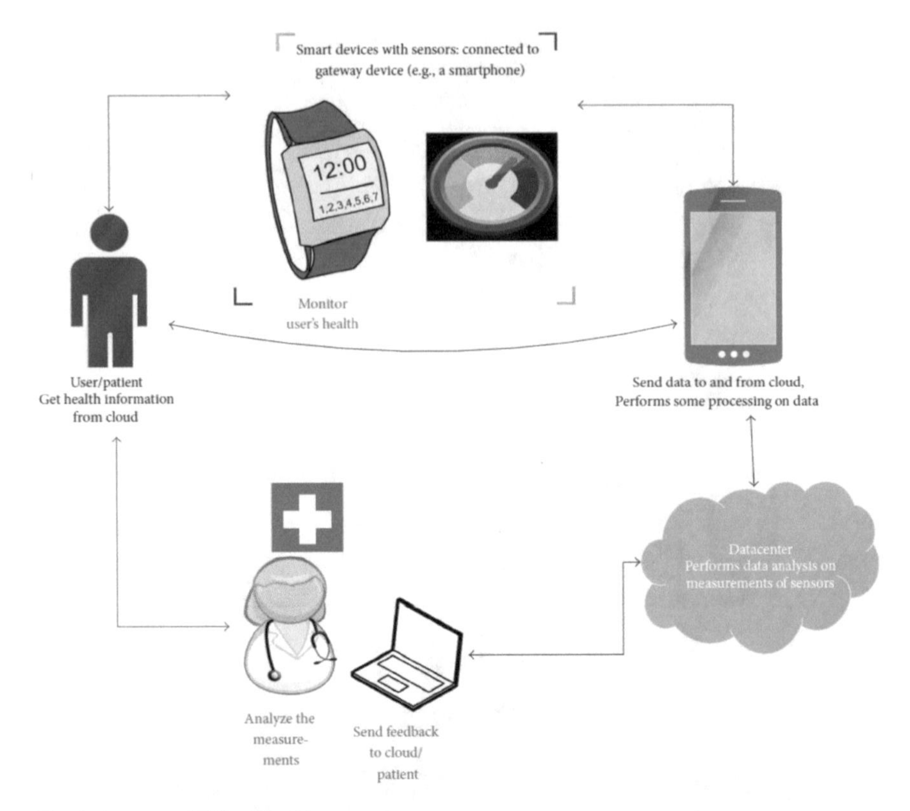

Fig. 3 Internet of Things healthcare system

and it sends the data through cloud platform and computes the data processing [19]. The data center is responsible for computing the measurements of the sensors and evaluating the performance, and the every detail is delivered to the doctor who is connected with this healthcare system [20]. The diagnosis report has been sent to the cloud system and the particular patient. The entire procedure is demonstrated with full detail in Fig. 3.

2 Methods and Materials

Security association between Internet of things and big data collection process: Things is a simple, an off-the-shelf, and a programmable device. Internet of things means that everything is connected to the Internet in Internet of things. Even animals, plants, and locations are connected to the Internet as well as all kinds of goods and products. The concept of Internet of things is same with that of ubiquitous communications—anywhere and any object—and assumes a connection to the Internet. In Internet of things, an infrastructure is created where information between people

Table 1 Characteristics of data generated in Internet of Things

Divisions	Characteristics
Data collection	Data collected by the program by hand not machine
Property of data	Data—much finer than existing data, generated from machinery, sensors, programs such as click stream, meter
Data owners	Data from outside organization where the production and management of data are not possible
Data type case	Unstructured data—user data, such as video stream, image, audio, social networks, sensor data, and application program data

and things and things and things is exchanged and communicated with each other when all things of the world are connected, based on IoT, via the Internet. Things are physical objects constituting the natural environment such as humans, vehicles, bridges, various electronic devices, eyeglasses, watches, clothing, cultural property, plants, and animals. In Internet of things, the characteristics of smart devices by mounting various sensors and communications capabilities on things in the current application environment of Internet of things are shown. Internet of things generates the data beyond the range of data processing within given cost and time in a system, a service, and an organization (or company). This feature is shown in Table 1 for the generation process of big data in Internet of things environment.

Data attributes side of the big data shown in Table 2 is structured or unstructured data that is too large, compared to existing data, to collect, store, retrieve, analyze, and visualize by existing methods and tools. It is also data beyond the range of data processing within given cost and time in a system, a service, and an organization. Figure 4 demonstrates the smart parking system that the mobile is connected with the data center and the availability of parking space is delivered to the gateway with cloud storage. The parking space with sensors is placed in the parking place to identify the parking places.

Even though big data has useful advantages as well as dangerous drawbacks, it will be actively used in marketing. It is information that has comprehensive consumer information such as gender, ages, hobbies, and interests. Analyzing big data does not

Table 2 Property of big data source

Variables	Data produced from computer	Data produced by human	Relationship data
Producer	Application server log	Twitter, blogs, e-mail, photos, bulletin board posts, etc.	Facebook, LinkedIn, etc.
Type	Orthopedic structured data stored in DB	Half-orthopedic Web document, metadata, sensor data, process control data, call detail data, etc.	A typical social data, documents, audio, video, images, etc.

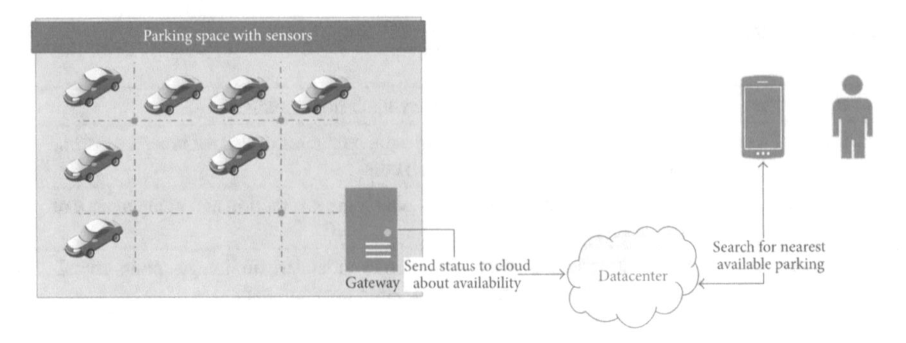

Fig. 4 Smart parking system

simply mean analyzing a large amount of data. System or service itself should have ability of adapting to big data as well as analyzing ability. Organization that plans, develops, and operates systems or services should have ability to deal with big data. Big data cannot be solved with one solution and should be solved with a variety of solutions, depending on the requirements and properties of data.

3 Results and Discussion

The Internet of things terminals hold the vulnerabilities itself. These are also same with the typical properties of information processing devices-hacker's attacks and malware infections. It can be seen that the attack patterns and threats threatening information security environment of Internet of things are equivalent to that of PC. Hackers tend to select attack targets that can generate a large amount of damage effect within a short time. Targets used to be devices which are popular and used extensively. Hackers attack or distribute malicious codes by utilizing OS vulnerabilities and protocol vulnerabilities discovered in smart TVs. Internet of things devices are small computers and of course use OS. By using Internet of things devices, hackers may do search, use social networks, and install new apps. Like security threats in the traditional database, data security threats in the Internet of things are serious threats. These can be threatening of spilling the backup of the entire file system with root privileges, spilling from insiders, and exporting the data by using DB administrator's privileges. The home automation system with the security, energy conservation, entertainment, and the lifestyle equipment with connected automation using the Internet of things is demonstrated in Fig. 5.

Internet of things may increase the analytical data, since traffic is increased and security threats are occurred by changes in various smart devices and the Internet environments. Data volume of Internet of things environment is based on size of the data. OLE data such as mail data or Web log data is correspondent to several PB; however, Twitter network data is less than several tens of GB. The analysis and

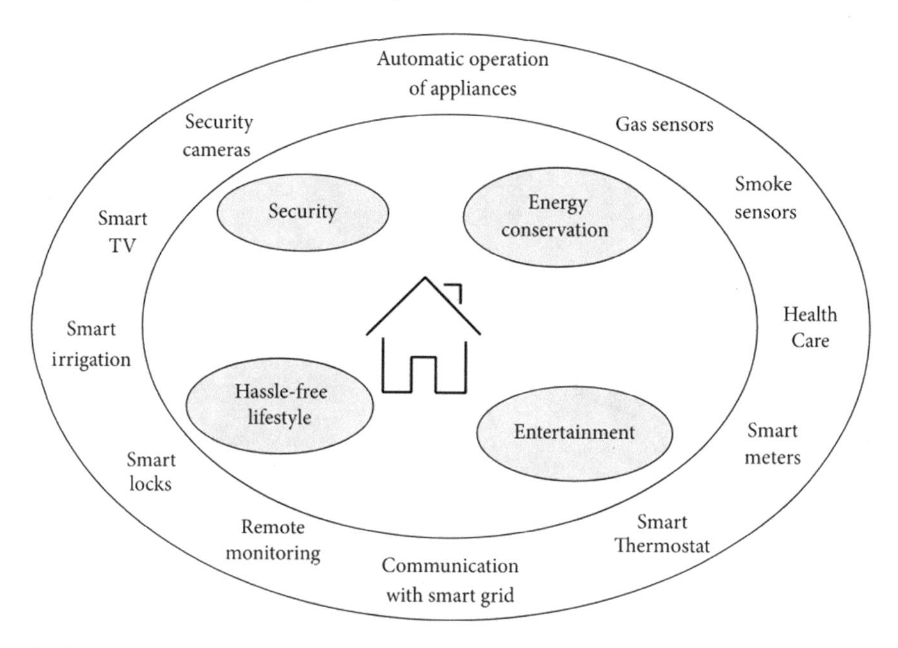

Fig. 5 Home automation system

processing of data are a significant concern. There is a difficulty in handling the attributes of data as shown in Table 3, since the properties of data, not the size of data, are important.

The velocity means the speed of data processing. It is a function that returns the result of processing after processing a number of user requests in real time, if necessary. The various data are analyzed in the traditional enterprise data and stored in ERP, SCM, MES, CRM, etc. It is the operating data generated from within the enterprise. Figure 6 demonstrates the application strategy and governance agenda which have connected the disrupters like cloud, big data, mobile and social networks using the application strategy of application modernization, and the data is placed

Table 3 Attributes of big data security quality

Divisions	Attributes
Confidentiality	Only the big data owner or a person who is authorized from the owner and a person who receive authorization from the relevant legal regulations can access to information
Integrity	Ensuring that the creation, modification, and deletion of big data information by an unauthorized party are prevented
Availability	The user is given permission to use big data information services at any time
Certification	Accessing to inside big data information assets from outside, it should be authenticated
Repudiation	Prevent to deny the action of the information systems usage

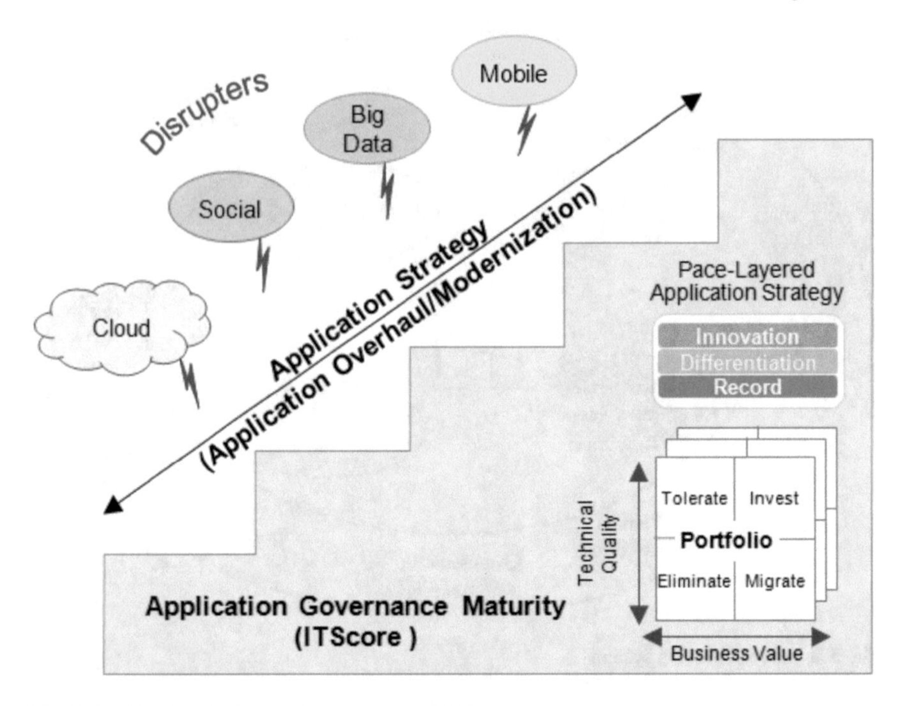

Fig. 6 Application strategy and governance agenda

in pace-layered application strategy with innovation, differentiation and record the entire process and the technical quality is checked through the portfolio like Tolerate, Eliminate, Migrate, and Invest. The application governance maturity having the IT score to gateway.

Some of the drawbacks are identified to connect the data through the Internet of things. They are

- Poorly managed and misplaced data.
- Absence of shared procedure.
- Missing standard approach for data standards.
- Unable to access old and lost information or if the information is there but not shared.
- Proper supervision of hardware media and backup devices outsourcing for backup and policies (potential threat).
- Complex scalability in existing structure.
- Unstructured data is not supported.
- Due to lack of any connectivity for unstructured data and social media, Company A was less response to the cases like Led Factor in children in Mt ISA, which is still in court.
- Lack of data quality and analysis on the datasets.
- Not proper data visualizations at high and global level to make decision
- Missing data connectivity through Oracle Warehouse Builder.

- Additional software required to connect through non-oracle data source.
- Inconstant data from multiple sources will bring data discrepancies and wrong decisions making, and data is getting bigger than ever before and doing analysis on large and massive is not that easy.
- To manage large datasets and analysis on massive data is time-consuming for financial transactions, which are getting bigger every day.
- Mobile assets and assets issues can be resolved through big data; mobile assets and assets management is the massive issue.
- RFID tracking is available on consignments and mine trucks but no information is stored and does analysis of RFID data.
- No real-time analysis on the business operations, which needs advance preparations for month end processes.
- Frauds in finance are paying heavy cost adding into Company A's budget value, which sometime requires PWC to find gaps between financial numbers
- Managed and low-level detailed information.
- Data analysis gathering requirements from process owners.
- Standard approach to follow data standards and business processes.
- HDFS will provide standard and proper backup and archival procedure which will replicate datasets, and there will be no additional investment will require on backups anymore.
- Scalability and availability of information.
- Migration data from different sources, structured and unstructured data to single data warehouse through HDFS and MapReduce.
- NoSQL will be used to gather all non-relevant DBMS information.
- Convert all the unstructured, media and Web data information into RDBMS.
- Information will be available all time for any sources and can avoid claim cases.
- Big data will provide more quality and detailed analysis for untouched data.
- Through big data advance analysis and global visualization for business owners not only for single site, which will also provide advance statistical and predictive modeling of the data.
- Provide the detailed analysis for planning, budgeting, and forecasting for finance subject area for better decision making.
- Need Oracle Data Integrator for big data connectivity adaptor, so all the information can easily integrate in the traditional warehouse system and ODI will also provide the data providers from reporting tool to the source level.
- High level of data accuracy, transparency, accountability, and consistency.
- To handle bigger and larger datasets big data appliance and tools needs to use to make more sense of data and detailed level of information.
- Handling large datasets is much faster and to make right decision at right time
- Provides advance analysis, data mining, and prognostic modeling techniques data discovery and explores different data patterns and associations to process and analysis on RFID data, and it is done only though big data which will support transport and logistics.
- Big data will provide you to do the analysis on real-time data rather than on historical data.

- Big data will provide the option in which decision or actions have associated risk involved with it.
- Also provide detailed analysis for the finance data to provide fraud management and facilities the business with data inconsistency to detect the frauds.

4 Conclusion

In order to cope with security vulnerability in Internet of things and big data environments, it is important to encrypt and authenticate messages sent between the nodes for establishing a secure wireless sensor network environment. An encryption algorithm and the key management protocol are necessary for encryption. The protection of privacy is also necessary in large data processing created at the same time on different channels. In order to meet the constraints and requirements of the security sensor network, a technique including a lightweight and password authentication technology suitable for environmental sensors, light key management techniques, and the privacy protection technology preventing side-channel attack and techniques must be used. The lightweight intrusion detection mechanism functions need to be applied to secure detecting node connected to the network. It is necessary to grasp the identification number of the terminal authentication techniques and systems for sensors utilized by humans. The increased traffic and security threats caused by changes in various smart devices and Internet environments, quality control programs procedures need to be institutionally and procedurally organized.

References

1. Bertolucci, J. (2018). Hadoop: From experiment to leading big data platform. *Information Week*. Retrieved on November 14, 2018.
2. Oussous, A., Benjelloun, F. Z., Ait Lahcen, A., & Belfkih, S. (2017). Big data technologies: A survey. *Journal of King Saud University—Computer and Information Science*.
3. Lakshminarayanan, K., Santhana Krishnan, R., Golden Julie, E., Harold Robinson, Y., Kumar, R., Son, L.H., et al. (2020). A new integrated approach based on the iterative super-resolution algorithm and expectation maximization for face hallucination. *Applied Science, 10*, 718. https://doi.org/10.3390/app10020718.
4. Sakaki, T., Okazaki, M., & Matsuo, Y. (2010). Earthquake shakes Twitter users: Realtime event detection by social sensors. In: *Proceedings of the 19th International Conference on World Wide Web, WWW 2010* (pp. 851–860). New York: ACM.
5. Kangin, D., Angelov, P., Iglesias, J. A., & Sanchis, A. (2015). Evolving classifier for big data. *Procedia Computer Science, 53*, 9–18. (Conference on Big Data 2015 Program San Francisco, CA, USA, 8–10 August 2015).
6. Balaji, S., Golden Julie, E., Harold Robinson, Y., Kumar, R., Thong, P. H., & Son, L. H. (2019). Design of a security-aware routing scheme in mobile ad-hoc network using repeated game model. *Computer Standards & Interfaces, 66*.
7. Ashish, T., Kapil, S., & Manju, B. (2018). Parallel bat algorithm-based clustering using mapreduce. In *Networking Communication and Data Knowledge Engineering* (pp. 73–82). Springer.

8. Shakarami, M., & Davoudkhani, I. F. (2016). Wide-area power system stabilizer design based on grey wolf optimization algorithm considering the time delay. *Electrical Power Systems Research, 133,* 149–159.
9. Santhana Krishnan, R., Golden Julie, E., Harold Robinson, Y., Kumar, R., Son, L. H., Tuan, T. A., et al. (2019). Modified zone based intrusion detection system for security enhancement in mobile ad-hoc networks. *Wireless Networks*, 1–15.
10. Harold Robinson, Y., Jeena Jacob, I., Golden Julie, E., Ebby Darney, P. (2019). Hadoop MapReduce and dynamic intelligent splitter for Efficient and Speed transmission of cloud-based video transforming. In *IEEE–3rd International Conference on Computing Methodologies and Communication (ICCMC)* (pp. 400–404). IEEE.
11. Paul, A., Ahmad, A., Rathore, M. M., & Jabbar, S. (2016). Smartbuddy: Dening human behaviors using big data analytics in social Internet of Things. *IEEE Wireless Communications, 23*(5), 6874.
12. Harold Robinson, Y., Golden Julie, E., Saravanan, K., Kumar, R., & Son, L. H. (2019). DRP: *Dynamic routing protocol in wireless sensor networks, wireless personal communications* (pp. 1–17). Springer.
13. Cappelletti, R., & Sastry, N. (2012). IARank: Ranking users on twitter in near real-time, based on their information amplification potential. In *International Conference on Social Informatics 2012* (pp. 70–77), Lausanne.
14. Poria, S., Cambria, E., Howard, N., Huang, G.-B., & Hussain, A. (2016). Fusing audio, visual and textual clues for sentiment analysis from multimodal content. *Neurocomputing, 174,* 50–59.
15. Hong, L., Dan, O., & Davison, B. D. (2011). Predicting popular messages in twitter. Paper presented at the Proceedings of the 20th international conference companion on World wide web.
16. Harold Robinson, Y., Santhana Krishnan. R., Golden Julie, E., Kumar, R., Son, L. H., &Thong, P. H. (2019). Neighbor knowledge-based rebroadcast algorithm for minimizing the Routing overhead in mobile ad-hoc networks. *Ad Hoc Networks, 93*, 1–13.
17. Hoeber, O., Hoeber, L., El Meseery, M., Odoh, K., & Gopi, R. (2016). Visual Twitter Analytics (Vista) Temporally changing sentiment and the discovery of emergent themes within sport event tweets. *Online Information Review, 40*(1), 25–41.
18. Ebrahimi, M., ShafieiBavani, E., Wong, R. K., Fong, S., & Fiaidhi, J. (2017). An adaptive meta-heuristic search for the Internet of things. *Future Generation Computer Systems, 76,* 486–494.
19. Gobioff, H., Ghemawat, S., & Leung, S.-T. (2003). The Google file system. In *Proceedings of 19th ACM Symposium on Operating Systems Principles (SOSP 2003),* New York, USA, October 2003.
20. Balaji, S., Harold Robinson, Y., & Golden Julie, E. (2019). GBMS: A new centralized graph based mirror system approach to prevent evaders for data handling with arithmetic coding in wireless sensor networks. *Ingénierie des Systèmes d'Information, 24*(5), 481–490.

Applicability of Industrial IoT in Diversified Sectors: Evolution, Applications and Challenges

Rohan Sethi, Bharat Bhushan, Nikhil Sharma, Raghvendra Kumar, and Ila Kaushik

Abstract Internet of Things (IoT) is defined as an interconnection among several devices which share data and all other relevant information across the network using unique identifiers without the involvement of humans. Any physical device can be easily operatable by using sensing devices, which reduces human workforce. The applications of IoT have been used in each and every field. With its ease of use of applicability in every sector, it is nowadays used in Industrial Internet of Things (IIoTs). In this paper, we present an overview of different emerging applications of IoT such as blockchain applications in Industry 4.0 and IIoT settings. As security is considered main aspect in any model; therefore, security metrics have been highlighted along with IoT-based architecture. Later section of paper describes the evolution of IIoT, revolution in industrial sector, collaboration of Industry 4.0 and IIoT, industrial automation, cybersecurity and data analytics, blockchain enrolled with IIoT, applications in public domain, etc.

Keywords Internet of Things · Industrial Internet of Things · Smart contracts · Security · Privacy · Blockchain · Industry 4.0

R. Sethi · N. Sharma
HMR Institute of Technology and Management, New Delhi, India
e-mail: rohan_sethi-22.98@ieee.org

N. Sharma
e-mail: nikhilsharma1694@gmail.com

B. Bhushan (✉)
School of Engineering and Technology, Sharda University, Noida, India
e-mail: bharat_bhushan1989@yahoo.com

R. Kumar
Department of Computer Science and Engineering, GIET University, Gunupur, India
e-mail: raghvendraagrawal7@gmail.com

I. Kaushik
Krishna Institute of Engineering and Technology, Ghaziabad, U.P., India
e-mail: ila.kaushik.8.10@gmail.com

R. Kumar et al. (eds.), *Multimedia Technologies in the Internet of Things Environment*, Studies in Big Data 79, https://doi.org/10.1007/978-981-15-7965-3_4

1 Introduction

In this modern industrial era, the technological advancements have revolutionized the vast but complex industrial environment to its utmost potential. According to the research, industries equipped with latest technological interventions which are perfectly aligned and pertinent in the industries tend to record a great statistical growth rate as compared to the industries which depend on the conventional techniques for their output [1]. Figure 1 shows market share of IIoT (in Million $) [2].

One of the most sought-after revolutionary technologies which could result in the expansion of the industries in their respective domains could be, 'Industrial Internet of Things'. It is believed to be a remarkable contributor in industrial systems of the future. Moreover, one of the greatest concerns for IIoT acquisition is 'cybersecurity'. Since, IIoT has been evolved from IoT, due to the indistinguishable architecture which is basic in foundation, which also relates and confirms the inheritance of certain security challenges from IoT [3]. But there exist certain differences among the two associative technologies which are expected to show great optimistic outcomes for the world. The main difference is observed only, in their areas of appropriate functionality, otherwise, the technologies are quite similar in their mode of operations. Some of the serious security challenges which IIoT face includes, monitoring sensitive information about the company records, confidential biometric information scanning [4] and there could be more, as in the field of technology unknown threats are bound to occur. Thus, the security concerns are classified into two broad groups: security issues related to the industries and security issues related to IoT and IIoT [5].

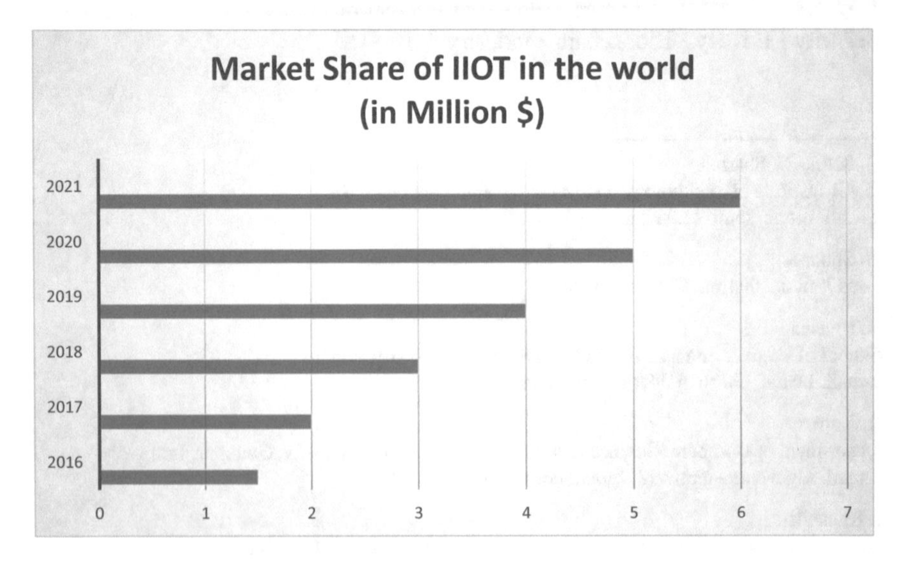

Fig. 1 Market share of IIoT (in Million $)

Industrial Internet of Things (IIoT), provides pervasive computing and networking abilities so as to introduce the industrial workflow with intelligence complemented with interconnection which is certainly possible with the devices such as sensors and actuators [6]. Since, industries consider security as their main concern; as it is a potential resource for the hackers to analyse the fruitful information of the industry their techniques of workflow and much more. And to perform their tasks effectively, 'cybersecurity' [7] forms a crucial aspect of the security concern for the industries. But still there are lot of points of discussion related to IIoT due to which the industries might consider this as a great alternative at least to those conventional techniques which demand high supply but provide low output. As compared to the successful installation and consumption of resources offered by IIoT devices in the industries [8].

However, IIoT benefits the industries in numerous ways, which involves automated and accurate exchange, analysis and data collection which ultimately leads to facilitation of advancements in efficiency and productivity which would lead to higher economic benefits [9]. Moreover, this technology focuses on the future scope as well, which deals with enormity of automation [10]. The automation adopted in industries may range from technology to technology. Here, the collaboration of cloud computing with IIoT is generally considered as an appealing factor for the industries. Since, cloud computing assists the analysed or raw data to be stored directly to the cloud [11] which is generally hosted by the industry. This procedure certainly leads to the refinement of the optimistic approach of IIoT in industries [12]. Also, IIoT is equipped with the latest technology interventions which in themselves are broad scenarios of the technologies to be considered which are: cybersecurity [13], RFID technology, Internet of Things, cloud computing, 3D printing, advanced robotics, edge computing [14], big data, mobile technologies and cognitive computing.

The rest of the paper is organized as follows. Evolution of Industrial Internet of Things has been discussed in Sect. 2, Sect. 3 illustrates the revolution in industrial sector, Sect. 4 shows the comparison of IIoT and IoT, Sect. 5 explains IIoT architecture, Sect. 6 highlights the collaboration of Industry 4.0 and IIoT. It draws security applicability in cybersecurity and data analytics, Sect. 7 formulates case study related to IIoT, along with applications of IIoT in business domain and combinations of blockchain technology with IIoT, Sect. 8 deals with the involvement of security challenges in IIoT followed by conclusion of the chapter in Sect. 9.

2 Evolution of Industrial Internet of Things

The metamorphosis of IoT from the consumer sector to that of the industrial sector in terms of IIoT was not tranquil. IIoT was one of the eminent contributors in the rise of 'Industry 4.0' as it is mentioned so as to distinguish the era of Industrial Revolution. Earlier, it was considered that Industry 3.0 could be the point of brim from where technical interventions could be at their peak to assist the machines and the industries at a broader prospect of production. But the revolutionary establishment

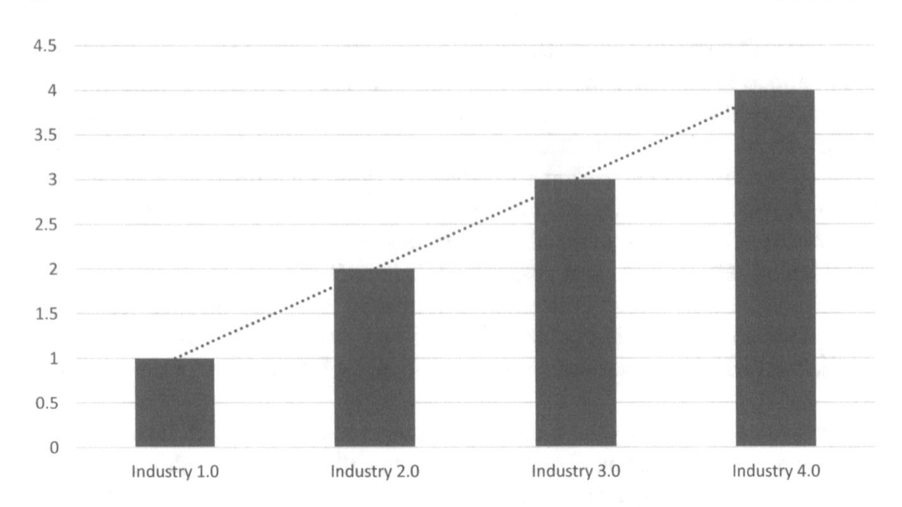

Fig. 2 Evolution of Industry 1.0—Industry 4.0

of technology such as IoT, cloud computing and many more, the list is endless; which made quite a remarkable stand in the technical sector and with their practicality left the world in awe. This led to the creation of seemingly every possibility which a human could imagine of, and implement them with tranquility which could transform the lives of mankind. The evolution of Industry 1.0–4.0 is shown in Fig. 2 [15].

IIoT systems are generally perceived as the layered modular architecture which are one of the significant components of digital revolutionary technology [16]. If we snoop its derivative technology, i.e. IoT then we could always have sense of potential of this technology, whether there would be its applications or ease of implementation at large scale. But nobody was that much aware or confident about its successful implementation in industrial sector in the form of IIoT [17]. IoT was earlier, used in establishment of the connection with devices at an experimental or at a very small scale, but IIoT has transformed this approach as well; by the aggregation from commercial sector to industrial sector [18]. Moreover, the other technologies moved hand in hand with IIoT to assist the proliferation of this technology to its utmost potential, including cloud computing, big data and many more. Although, IoT targeted the connection of wired and wireless hardware devices [19], but still the technology lacked the potential to demonstrate such interventions at a large scale, and this paved the way for the rise of IIoT in the industries [20]. Since, IIoT targeted the convoluted physical machinery which is associated with actuators, sensors and other related software applications which could ultimately drive through the change of workflow in the industrial sector [21].

The main targets of any industry includes high growth rate, high efficiency, high profit, high quality with reduced overhead [22] and that is desirable in association with low cost and input. This led to the emergence of adoption of this technology in the industrial sector which could advance the conventional and other so-called modern techniques which require a large amount of effort both in terms of capital invested and the input supplied [23]. The vision of IIoT is simple and crystal clear and, that is, to establish every enterprise as a smart production unit which could easily connect with the systems at a larger scale and contribute in the productive formulation and a strong foundation of the industrial sector worldwide [24].

3 Revolution in Industrial Sector

The track to the development of Industrial Internet of Things (IIoT) initiated much early and that was in the year 1968, when the term was first coined in the technological and industrial domain by Dick Marley, which was considered as one of the greatest breaks through in twentieth century. But this major breakthrough came in assistance with his other companions, as they invented the 'programmable logic controller' (PLC) which led to the automation of numerous industrial robots and assembly lines [25], which proved to be one of the most stupendous art of revolution in the industries. This was impossible without the developments and interventions in Internet of Things (IoT) as this technology was responsible to a greater extent to provide all the basic resources and the authority for Industrial Internet of Things (IIoT) [26]. This led to the urgent ratification of automation in the industrial sector, which led to the introduction of conventional 'single board computers' (SBCs) and I/O components to be utilized progressively in the manufacturing process throughout [27]. The Industrial Internet of Things' (IIoT) acts as one of the parts and parcel of the 'Industry 4.0' which targets greater safety, greater efficiency, reduced overhead and high quality [28]. IIoT supports rapid progression for the rise of Industry 4.0 [29], which is setting a benchmark for the application of business processes in the industrial sector. It has initiated the process of transforming the industrial sector altogether, in every astute manner, moreover, the grandiose automations in the industries demonstrates a clear sign of the evolutionary presence of IIoT in the origins of the functionalities and technologies in the industries. Figure 3 shows global shares of IIoT [30].

The outcomes of the proliferation of this technology in the industrial sector have provided a way to the machine builders and the end users to grapple the prevailing investment in people and linked technologies, while taking full benefits of the accessible technologies. But there still prevails the utmost confusion among the investors and stakeholders over, the operation of such technologies in the industrial sector, since most of them target and analyse the impact of this revolution on the automation sector of the industries. The vision of IIoT is relatively intuitive in the sense, of promotion of adoption of automated methodology in the industries and be operative as the part of huge systems which together focusses on building up smart manufacturing enterprise [31].

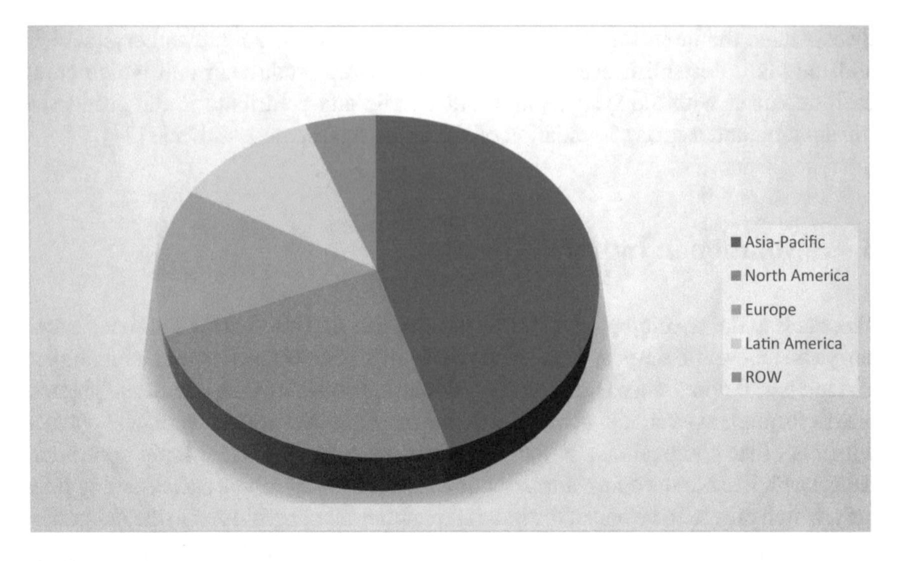

Fig. 3 Global share of IIoT

4 IIoT via IoT

Internet of Things (IoT) is a completely different technology when we consider the applicability, as it operates and functions in the consumer sector while Industrial Internet of Things (IIoT) operates and functions in the industrial sector. Internet of Things (IoT) is that form of technology which deals with the connection of physical devices through wireless and wired network, whereas Industrial Internet of Things (IIoT) is that form of technology which is accorded with the convoluted physical machinery associated with the industrial actuators and sensors with the suitable software applications which could provide the workers an interface to interact with the hardware and other physical machinery to proper effect as required to accomplish suitable manufacturing needs. IIoT involves an accurate and precise approach for the provision of connection between the machines and humans.

Since, the application of conventional techniques in the industries for various purposes including production assembly lines, planning, quality control and administration, the industries always experienced low graded terms in all these factors discussed. This might be due to the fact that, most of these techniques involve much labour, high input in terms of capital invested, materials used for the production and the manufacturing process followed by packaging and delivery on the basis of supply and demand; the lack of automation consistently questioned the approach with lower profits of the investors and stakeholders. But with the advent of Industrial Internet of Things (IIoT) paved the way to automation, analysis [32], control and monitor of the processes involved in the industrial sector and that also with sheer accuracy and precision; due to which the investors and stakeholders found the way to capitalize their investments with increased profits and benefits to both the industries as well as the end users. So, due to these major determining factors, Industry 4.0 could well be assisted with Industrial Internet of Things (IIoT) to pave the way to totally evolve the way with which workflow is borne in the industries and to reinstate the conventional techniques which are worn out and are more putative. Moreover, according to the reports, the overall functionality operation and demonstration of the adoption of Industry 4.0 by the industries are not considered atrocious, in fact, it could stimulate the growth and sustenance of R&D and IT sectors.

This could also lead to the ascent of various emerging technological models in industries such as on-the-spot 3D printing, machines-as-a-service and many more which focuses on these technological domains. Since, this applicability of IIoT augments the utility of entirely new and in-demand business models and product offers. In 2010, this complex yet amazing technology was given an initial boost and platform by Google, when this tech giant delivered a breakthrough in this technology domain by the demonstration of their developed 'Google Street View'. With the advent of which, the 'Government of China' also proposed this revolutionary technology in their five-year plan, with the vision of future endeavours to accomplished by the country with this technology. Subsequently, the 'Government of Germany' provided a thrust to the revolution of IIoT. Figure 4 shows the progress of Google Street View for business [33].

However, IoT devices are considered as those devices which just switches on/off remotely, but:

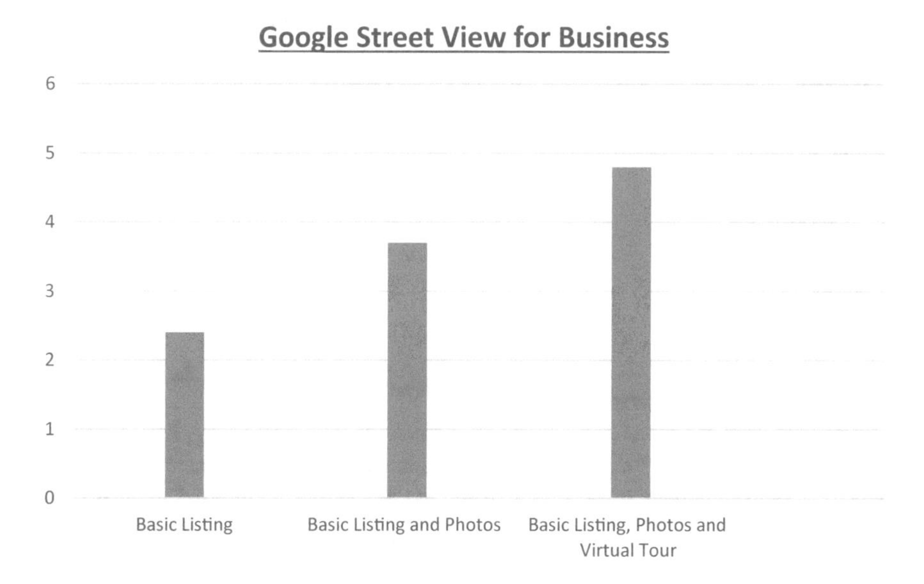

Fig. 4 Progress of Google Street View for business

- Smart devices accompanied with their self-processing units.
- Possess numerous techniques which exchanges vast amount of information with each other.
- Can control the entire network, transmit and receive the information as required.

Although IIOT is considered as the technology which has evolved from IoT, but IIoT transforms such functionalities and takes it to the next level, because it caters a standard and a system for the interconnection at a global scale through Internet Protocol(IP) [34]. The Internet Protocol or IP, the term with which it is better known as, is actually a unique identifier which swaps the information or data signals over the inter-network or better known as Internet, which is independent of the devices between which information is shared in exchange. And, by utilizing the similar architecture and protocols, the communication is successfully established with other devices as well, which are on the similar network. Internet Protocol (IP) [35] establishes a crucial pathway of interconnection between physical devices along with the well-established network infrastructure.

5 IIoT Architecture

Industrial Internet of Things involves the interventions of big data and machine learning technologies in association with supervisory control and data acquisition (SCADA) and programmable logic control (PLC), which focusses on the improvisation of automation of technology by implementation of rectification ability and self-diagnosis. However, in the industries which are controlled by the well-established procedures or processes, the physical devices involved are dependent on the physical factors such as:

- Pressure
- Temperature
- Vibration
- Flow
- Level.

And numerous other factors apart from the discussed above are also responsible for the same.

These factors are generally helpful in the collection of data only, but as per the current scenario, they are also convenient in exchange, correction and analysis of the data in an entirely advanced way. According to the analysis by the IIoT experts [36], these technologies would definitely assist in the improvement of efficiency, green practices, sustainability and supply chain management. The shift in the trends of IoT, to the industrial level from a foundational commercial level, is responsible for the influence in terms of continuity and efficiency of an industry universally. IIoT adorned by the idea and methodology of Internet of Things, to the operational level that of the industry. Consequently, each enterprise acquires the exclusive collection of devices which experiences the limitation in terms of interfaces. On consideration of challenges to be faced, there does not exist any solution which could be able to solve the entire set of problems [37]. The major key components of the architecture of IIoT can be explained in detail as follows:

- Industrial Control System: The industrial control system can be defined as the generic term which is to explain the hardware and software integration which could be able to control the critical infrastructure. Such technical practices generally involve the development using human–machine interface (HMI), supervisory control and data acquisition (SCADA) system, distributed control system (DCS), programmable logic control (PLC), intelligent electronic device (IED), remote terminal units (RTU) and many other relevant industry-specific systems [38].
- Devices: Some of the industry-specific devices, includes interpreters, sensors, translators which involves the interfacing with transient data stores, ICS, processors and channels which avails the necessary data to the application at the user end. They are involved in human-to-machine synergy, machine-to-machine synergy and vice-versa which enables the competence to industrial control system.
- Transient Store: This key component is also well known as the slave component of the master architecture which involves the ephemeral depiction of the objects involving data and is stored for a short period of time, also better known as data store objects. This assures the endurance during the system failure and operation failure, which involves networks.
- Channels: They can be explained as, the medium for data exchange between the application and the system. It involves satellite communications, routers, API, network protocols, etc.
- Processors: They are also known as the heart of IIoT solutions. Their main functionalities involve signal detection, complex event processing, analytical models, data transformations, etc.
- Models: The two models which are comprised in the component of models in IIoT are, 'data model' and 'analytical model'. The data models explain the structure of the data while the analytical model follows the industry-specific needs which contributes the custom build [39].
- Collectors: The collectors can be defined as the data assembler, which gathers data by the usage of standard protocols. These devices are independent in nature, as they vary from industry to industry.
- Security: This is one of the most important aspects of the systems which are based on IIoT. It works on the basis of pipelines which originate from the source to the destination where, at the destination, the ultimate consumption of resources takes place. It involves firewalls, encryption, user management, masking, data authorization, etc.
- Gateways: They provide the connection across numerous protocols and networks which enable transfer of data between various IIoT devices. It comprises of information transfer protocol, intelligent signal routers, etc. [40].
- Permanent Data Store: They can be defined as the data storage devices, which involves data storage for a long duration of time, connected to the IIoT system.

It comprised a huge amount of open-source data, cloud storage, RDBMS, data repositories, parallel processing system, parallel processing data stores, etc. [41].

- Local Processors: The local processors can be defined as the low latency processors, which supports rapid processing of data and can be combined for the purpose of processing the data with the device itself.
- Fog Computing: The technology proposes the analytics adjacent to the origin.
- Application: The component which provides the intuition to the on-field operations in real time, as they assist the staff members to manipulate the data after successful interaction with the system. The effective and calculated decisions are ensured due to better visualization, notifications and alerts.
- Hybrid Computing: The combination of cloud and fog computing which optimizes the operations which are created and developed specifically for targeted fields in need.
- Cloud Computing: The technology which scales analytics on a global scale across the industrial sector.
- Computing Environment: The environments discussed are independent of industry which implies that they are different for different industry which depends upon the landscape and its business.

6 Collaboration of Industry 4.0 and IIoT

The movement of revolution in the industrial sector was triggered initially by the 'Government of Germany' in 2010, due to which the country saw the development of the manufacturing sectors where cut-throat competition among the related sectors is observed. This ultimate adoption of the revolutionary step in the industrial sector was considered as futile by other countries, but the 'German Government' was sanguine about their trigger movement, and then what the world observed was history, as Germany became the global leader in the industrial sector of production of the necessary equipment's. The major target of Industry 4.0 is well clear with the vision of this technological revolution, which aims to obtain smart, productive and high-quality adoption of the relevant processes for the purpose, which also focus on the introduction of those techniques, which increases the growth and to accomplish those results which were considered impossible, a decade ago. The Industry 4.0, not only involves the ratification of conventional techniques but also are focussed to automate those techniques to save high productive time, cost and labour. Figure 5 shows the impact of Industry 4.0 in the world [42].

The industries have consistently been involved in the practices which leads to the increase in production, profits and trade. And with the rise of such technologies, such as IoT, IIoT and Industry 4.0, the industries experience a merry time as these technologies not only supports the end users with high quality, low prices, etc. but also they benefits the industries with increase in their profits, low input cost and better techniques of monitoring the practices involved in the industries. The Industry 4.0 constitutes of industrial and conventional manufacturing practices with the increase

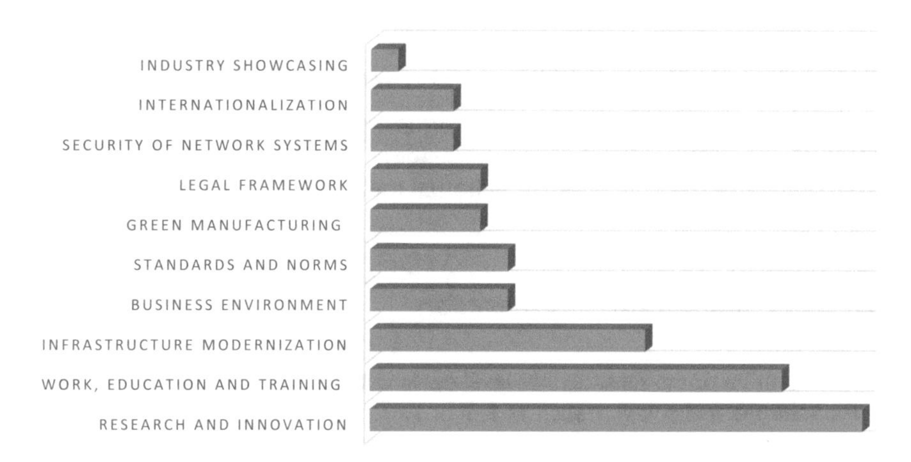

Fig. 5 Impact of Industry 4.0 in the world

in technological interventions around the world. This comprises of using Internet of Things (IoT) and machine-to-machine embroilment which supports the consumers and manufacturers in the similar manner, with the developed communication, monitoring and increased automation, which involves self-analysis and diagnosis for the advanced levels of analytical procedures to support a great constructive feature [43]. Industry 4.0 involves communication technology and broad information which introduces the evolution in the production and supply chain management with ultimate digitization. It connotates artificial intelligence to complete tasks, self-optimization and self-configuration in the machinery which results in better quality of goods and services and cost efficiencies. According to the industry experts, due to the above-mentioned notions and interests, the industries opted to adopt the initiation of the evolution of IoT to a larger scale and to an entirely advanced era of technology, which is better known as IIoT.

The definition of IIoT according to the market analysts around the globe has explained the concept of IIoT as mentioned below:

- The Industrial Internet of Things (IIoT) is known as the successive step of modernization which leads to the clout for the path through which the world connects and improves the machinery of the industries.
- The O&G companies which are the world leaders in this sector are currently focussing according to the circumstances to build an optimized framework which involves advanced analytics, sensors, automation and data management to focus more on the production to amend the assets of the industries and to curtail the operational costs.
- The Internet used within the industries can be defined as the way in which the intelligent machines work and the procedure with which they are designed and would transform the business involved in the industrial sectors, similar to the way in which the Internet moulded the lives of the consumers [44]. And this form of

Internet in the industries which involves interconnection of machines and other components is known as 'The Industrial Internet'.

Since then the industries along with the other companies across the globe had struggled the way through to the implementation of IIoT in the manufacturing sector to be witnessed as the component of Industrial Revolution.

6.1 IIoT and Industrial Automation

The distribution of the sensors, low-level devices and actuators which are required to facilitate the technologies such as IoT or IIoT, due to which the automation of industries could well be benefitted. Most of the industries in the industrial sector are patiently pondering over the current circumstances and waiting for the distribution of the connected devices at a low-level to initiate the operation of IoT in industries.

The transformation of IoT to IIoT has been possible due to various outcomes as well, one of the determining factors is to consider a well-established platform for IoT to perform and function well in the commercial sector to evolve in the application domain of the industrial sector. Due to this effective functionality of IoT the radical and contemporary technology of Industry 4.0, which is known as the Industrial Internet of Things (IIoT). The improvization of industrial network their operation and utilization is one of the key applications of Industrial Internet of Things (IIoT) in terms of automation of the whole network [45], both in the future and under current scenario. The distribution of the end devices which are used in the connection within the industries, generally known as 'Intra-net' and connection with other industries, which is known as 'Internet'; on the scrutiny of data which is a valuable asset of twenty-first century, which is noticeably accessible at the end devices and the various cybersecurity techniques. IIoT safeguards the industrial systems from downtime, as they involve the latest technological interventions in the industrial sector, which proves to be effective as they detect early signs of defects or malfunctions which could ruin the industrial sector and adopts the prevention policies against the aggravation [46].

The installation of IoT sensors and their introduction in the industrial sector permits the operators in the industry to analyse the operations of the sensors and actuators installed in real time, with high precision and accuracy. Due to the prediction of the problems, which is a process carried out much early due to IIoT, prevents the lockdown operation, which generally means the state when the industries are non-operative. According to the discussion of the contribution of IIoT in the industries, it caters to whole lot of operations effective to the industrial sector in terms of performance, speed of response, security of the system and accuracy, which is responsible for the transformation of IIoT to the next level.

6.2 Cybersecurity and Data Analytics in IIoT

The term 'cybersecurity' can be defined as the technology which provides a shield to the connected systems which includes software, hardware and the protection of data from ensuring safety against the malpractices such as hacking, cracking and much more, which could prove to be a threat for the industrial sector. The security is the major concern, whether it is related to the data, devices or operations which are implemented in the industries with appropriate guidelines to achieve effective functionalities. These technologies are co-related with the security measures which are involved in both the linked technologies from both ends, for the security of the technological domains [47]. The technologies such as IoT and IIoT must also ensure they adhere to the effective security measures, so that the consumer as well as the industries could trust on them and could utilize their resources without any risk of the lack of security in terms of the provision of data related to storage, operation and activity control.

The field of 'data analytics' proves to be one of the effective tools to determine the statistics of productivity, to improve them or improvise the current crucial circumstances to the utmost extent. In the modern industry, the field of 'data analytics' advances the efficiency of operations which are executed in the industries. Since, 'data' is considered as the fuel of twenty-first century, and the domain of 'big data' is one of the trending technologies in which most of the investors and stakeholders are interested in providing the capitals, investments and other resources which could boost their businesses, industries and ultimately could also result in effectively boosting the nation's economy which could result in drawing the attention of foreign investors and stakeholders. Figure 6 shows the benefits of industrial analytics [48]. Such technologies are the need of the hour to sustain the efficiency

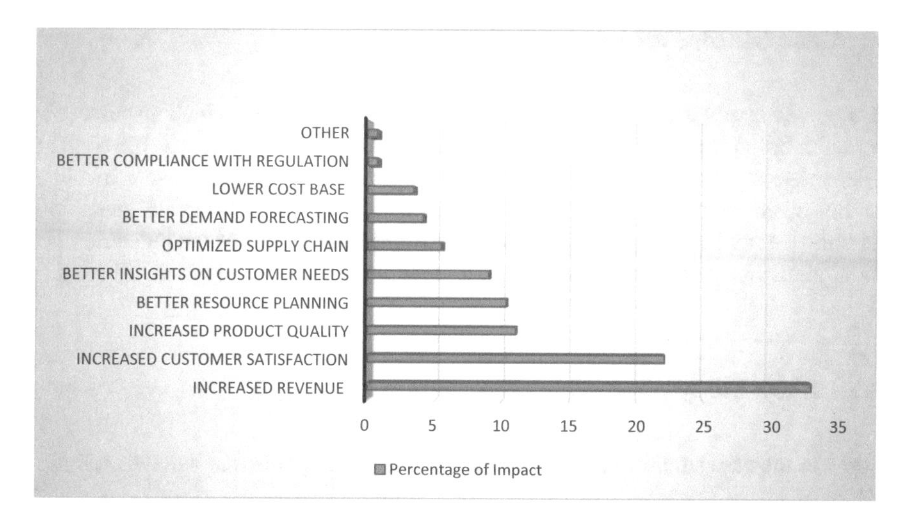

Fig. 6 Benefits of industrial analytics

of this global industrial sector. The categorization of big data in the technological domain is mentioned as follows:

- Data storage
- Data security
- Data analysis.

In the technological domain of 'Industrial Internet of Things', the technique of analytics is further categorized in various forms, such as business intelligence level analytics, massive analytics, memory level analytics, real-time analytics and offline analytics. The further operations are accomplished with the different forms of data, i.e. the analysis with audio, video, images and the sensor data [49]. The branches of the technological field of data analysis [50] are as follows.

- Cleansing: According to the data cleansing process, the data which is corrupted can be deleted, replace or modified in any form as required. And through the process of scripting, the process of data wrangling could easily be performed and implemented. The data which is resultant to various revamped processes must be free from errors and defects to result in effective processing.
- Modelling Data: In this process, the creation of a data model, for the knowledge and information is developed after the successful application of formal techniques.
- Inspecting: In the process of inspection of the transformed data, the depraved or faulty record, determined from the database and the data is transmitted for subsequent operation.
- Transforming: According to the process of data transformation, the data is altered from one format to another. After the successful termination of this process, the process of data modelling is initiated.

7 Case Studies and Applications

Since, we already are familiar with the concept of case studies, which discusses the research-based strategy after the successful execution of those, in the real-life context. These case studies provide a clear overview of the strategies which were involved in the consideration of the policies and guidelines which an industry followed with an utmost extent sufficiently, later after the disclosed strategies of the industries, the other industries could also follow those to be more productive and more efficient.

7.1 Case Study

Some of the case studies discussed in this chapter are explained in detail as follows.

7.1.1 Black and Decker

The well-renowned industry, which is a well-established manufacturer of power tools which has transformed into Cisco to implement the strategy of IIoT-based solutions for escalation of complexity in the industrial sector, when the intricacy is observed in its manufacturing unit in Reynosa and the availability of discernible wireless connectivity. Organization's Wi-Fi compatible programmable logic controller modules which are assimilated in association of Aeroscout's Wi-Fi tags. Actually, when the product arrives in the goal line, then these units are operated to monitor the specific quality criteria. The production plant acquires a growth in the rate of around 10 in terms of labour adaptability and other analytical resources. It enhances the overall application of rates up to 90%. There is also a reduction by 16%, in the quality defects per million opportunities.

7.1.2 AW, North Carolina

The company which is one of the leading manufacturers in the elements of automatic transmission of various categories in AW North Carolina. The improvisation of manufacturing plant efficiency is achieved by implementing the technology in a smart way. The area of one of the largest factories of AWNC is located in Durham, England is around 1.3 million square foot. Such a large factory must involve employment at a larger scale to sustain the maintenance of the factory and due to this purpose and functionality there are around 2000 employees currently employed in the factory. Moreover, the efficiency of employees is measured in transmissions per year, which also serves as the determinant of the rate of growth for an industry, and in general, the rate of transmission for this factory is around 600,000 transmissions per year. And according to the reports of analysis of AWNC, there are around 700–800 parts in each unit which are immensely specialized. Every part consists of definite structure of its own.

7.1.3 Hirotec

The Japanese manufacturer which is renowned as one of the enormous private manufacturing organizations in the world in market segment of automation. The company planned and scheduled the implementation and adoption of Industry 4.0 techniques and methodologies to initiate the innovation for the purpose of capricious downtime. This manufacturing group planned to break through the functional capabilities assisted with predictive analytics. One of the prime concerns for every industry in the industrial sector is to reduce the occurrence of downtime in the industries, which also involves prevention of downtime in the manufacturing facilities, within any organization. Due to the introduction of Industry 4.0 and the proposition of the related manufacturing and production models in the market, it serves as a major contribution in and critical for all the industries to accomplish those results which were effectively accomplished by Hirotec efficiently.

The Hirotec company demonstrated the eminent contributions by the three pilots of the company. They acted with a unique but effective approach as they initially analysed the data which was preceded by the recording of the data effectively where CNC machines performed the role of source aggregators, the implementation was accomplished in Detroit plant of the company. The company enforced the platform of IIOT, in the successive pilot mode, which was initiated with an aim to function the remote observation of the exhaust systems in automated version. The discussed purpose was operated with the cooperation of various data sources, such as sensors, cameras, laser-based measurement equipment's and inspection robots. The company initiated the incorporation of the systems to achieve observation in real time associated with the full production lines of the automobile systems which is assisted with the facility of automated report generation.

7.2 Applications of IIoT in Business Domain

According to various tempestuous and eminent applications, the monitoring maintenance in predictive form is one of the critical applications of IIoT in the industrial sector. It consists of numerous sensor data to a limit which is practically possible to a certain limit, which includes, humidity, temperature, current, density, vibration, voltage, etc. by applying the effective algorithms of one of the best technologies, machine learning. Some of the efficient algorithms of machine learning can be able to anticipate the chances or probability of failure at an early stage. In the industrial sector or elsewhere, the companies apart from increasing the quality, efficiency and reduction in costs and inputs, focus on diminishing the zero safety incidents, environmental incidents, number of accidents and zero break downs [51]. The availability and installation of sensors in any machine could well be able to validate the health data points of machines and issue warnings accordingly, but there is a lack in the functionality of such devices as these devices are not capable enough to determine the reason of their failure or inability to perform the required function accordingly [52].

The discussion for the concept of maintenance is simply to formulate a system which could be responsible for administering the meticulous probability indications on the data, comparatively functional for only addressing it. Let us consider an example, the automated system in the industrial sector is so efficient and effective that it could control and monitor the whole industry at once. The system is capable to anticipate the failure of a component, for the involvement of all the components of the industrial sector and intact the efficiency of the whole production unit, and the automated system assigns a time period within which the replacement of the faulty parts or the components should be replaced to maintain the efficiency and places the order for the defected or faulty component well in advance so that the ordered element is delivered in time.

The proper functionality is achieved after the execution of above-mentioned strategy, and highly efficient with the constructive output of high-cost efficiency. According to the reports, after the effective analysis by the industrial experts, it is claimed that from an IIoT perspective 'transportation' is the second-largest market in the global industrial sector [53]. With the advent of monitoring system and technical communication, the value chains system is entirely based on IIoT on in-depth consideration of logistics and transportation firms [54]. Hence, it proves that the impact of IIoT in the field of automation and industrial revolution is so effective across the globe that it has entirely transformed the automation sector of the industries [55].

7.3 *Collaboration of Blockchain with IIoT*

The looming and one of the most sought-after technologies in the technical domain of twenty-first century are the blockchain technology which depicts a promising capability for the enhancement of the Internet of Things (IoT) and the industrial machinery and systems by contributing for the applications with superfluous, encryption and immutable storage. In the past, the technology of IIoT has appealed to the academic scholars as well as the researchers, due to the wide variety of applications in the industrial domain. And, at the same time, the technology of blockchain also loomed parallelly which was also responsible for drawing attention of the researchers and academic scholars to progress in this path of research, which is based completely on significant contributions from their end. The association of IIoT and blockchain serves as one of the critical collaborations together in terms of industrial perspective [56]. The fundamental techniques which lead to the creation of IIoT framework which is well established and operated with the assistance of blockchain technology to address the key aspects and challenges faced by these technologies together are addressed effectively. Figure 7 shows the rise in the global market share of blockchain (in Billion US Dollars) [57].

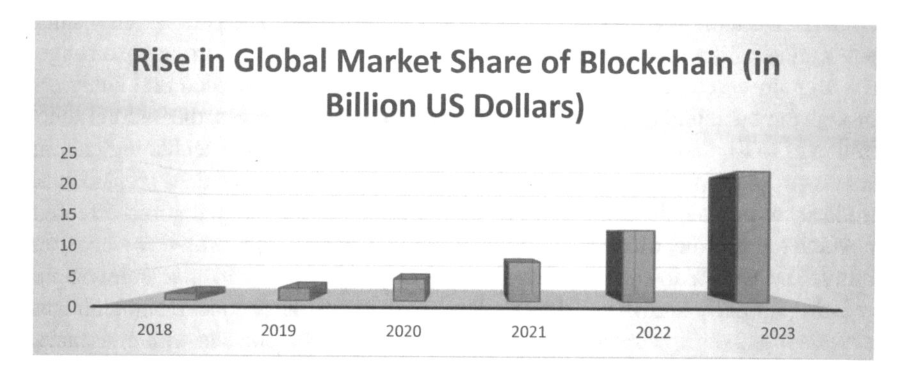

Fig. 7 Rise in the global market share of blockchain

On the most recent successful contribution of exhaustive analysis for the purpose of open issues and research trends is introduced with the IIoT assisted with blockchain effectively. The technologies which have been emerging under current circumstances, which includes Internet of Things (IoT), Industrial Internet of Things (IIoT) and many more, which has entirely transformed the way in which the technology was perceived a decade ago, it has also set up the creation of the framework which is responsible for a significant increase in the research and discovering new themes of research which involves the industrial application [58]. In IIoT enabled systems, the IIoT platform performs a critical role which could deliver throb operations of connectivity, which involves assets of connectivity and enable IIoT with various abilities such as, big data analytic, application development and connectivity. To search for the solutions of the above-mentioned challenges and problems, there are certain requirements which IIoT platform must contemplate:

- Ageing workforce
- Integration of technologies
- Data islands
- Cybersecurity
- Visibility of asset.

Moreover, the customer's or user's behaviour is also often dependent on IIoT. As most of the present facilities in the industrial sector which includes smart-grid IoT, micro-grids, vehicular ad hoc networks (VANETs) and so on, which are created to connect, but they are unable to show any terms of their connections to IIoT in assistance with intelligence which already exists and that depends on the need of interfacing to establish communication. Moreover, the assistance with newly developed technologies, such as 'augmented reality' which could lead to a better forecasting process behaviours and interaction and could result in improvised efficiency and uncomplicated functionality in the industries [59].

The first platform that transformed the technology of blockchain was 'Bitcoin' which led to determination and depiction of the utmost potential of this technology in general; which focussed and targeted to greater efficient processes to initiate and also led to the creation of a path which was not at all expected. The path which opened new avenues, like that of reliable, traceable and cheap cryptocurrency exchange. The Bitcoin-based transactional activities could easily be recorded and analysed, through the operation of only IIoT devices which further employs, the internal functioning of IoT-based techniques. The widely used platform in IoT technology is 'the Ethereum' platform which is enabled via 'Ethereum virtual machine' (EVM) used primarily to provide flexible consensus strategy and built-in-smart contract features, in which the contract offers smart operational IIoT applications with down compatibility. The well-renowned blockchain technology platform which was developed by IBM popularly known as 'Hyperledger' is an open-source blockchain platform. It offers membership strategies categorized industrial components with consensus. Moreover, it known to provide a full-fledged support to 'IBM Watson IoT Platforms'. The Hyperledger platform is observed to contribute in processing speed of IoT applications.

The other developed platforms in blockchain technology specifically for IIoT includes the 'HDAC', 'Litecoin', 'Multichain' and 'Quorum' which are capable to provide the applications for IIoT with trustworthiness, traceability and many more. The measurement of the platforms could be achieved through the cup utilization, the size of block, energy consumption and many more.

8 Security Challenges Involved in IIoT

This section deals with the security challenges, which are featured in the environment of IoT and also sustains in the scenario of IIoT which analyses the conforming solutions [60].

- **Data Confidentiality**: The applications which connect the machines, actuators and sensors in high-stake industries which will assemble a huge of volume of non-uniform data which needs to be refined in real time. At some instants, the factories, industries or even the critical infrastructures could result in calamitous state for the industries. The process of data encryption in data and network security involves the conversion of data in ciphertext which involves the transmission and storage and proves to be one of the most common techniques to safeguard the confidentiality of data.
- **Cyber-Physical Systems Integrity**: The most of the IIoT and IoT devices are considered as CPS and the integral verification of CPS systems should be well supported in the IIoT and IoT systems. An eminent mechanism, which verifies the integrity of a software configuration in the system, also known as attestation, which is responsible to facilitate the detection of malicious and unintentional software modifications. The fundamental concept behind the application of attestation and verification for integrity is possible with the help or assistance of a device called prover, which on its application sends the report in response which provides the status in the form of attestation and the software configuration for the current scenario is transferred to another device known as verifier; the verifier demonstrates that it is a trustworthy state and a well-known state for the verifier.
- **Pivotal Establishment for Pairing of Devices**: The IoT/IIoT devices which are deployed recently must be able to pair securely with the other existing devices and that is possible through cryptographic key development. The protocols which deals with the development of asymmetric keys involves such devices which are required to compute costly operations, this includes the modular exponentiation operations, which sometimes creates an uncomfortable situation for the IoT/IIoT devices which are limited in terms of resources. Whereas, the symmetric key establishment protocols need such devices which are pre-loaded with the secrets which are shared and the keys can be developed dependent on the basis of shared secrets.

- **Device Management**: With the advent of IoT/IIoT in the technological domain associated with industrial domain, apart from the rise of the technology parallelly the rise in the numerous challenges linked with the field are also numerous, due to which in the current scenario, one of the greatest challenges in the field is to adopt the procedure to securely and efficiently monitor large deployments of devices in the industries [61]. The process of device management must assist the tracking and governance with appropriate control at every subsequent lifecycle of an IoT/IIoT device which includes the process to execution results, monitoring status, updating firmware updates, sending commands and resolving the problems interactively.

Hence, the impressive and efficient management of IoT devices is crucial and fundamental for building compelling systems in the industries.

9 Conclusion

The Industrial Internet of Things and Internet of Things are capable to lead the transformation and could also result in the determination of colossal power which could provide numerous applications across industrial domains and disparate commercials. This review provides an application roadmap and a research with its extended vision in a broader sense. All the systems of the industry are required to be sensitive to a higher degree regarding the security and safety of the IIoT devices to avert the harm to the personnel and the assets. But still there exists a big question mark on the security of IIoT devices, since security is the major concern of every industry in the industrial sector, as the data is critical for the industries as well as the consumers. The detailed research work is required at greater extent, which could possibly solve such issues related to the security which are prone to besiege the field of IIoT and its implications in future.

References

1. Puthal, D., Malik, N., Mohanty, S. P., Kougianos, E., & Das, G. (2018). Everything you wanted to know about the blockchain: Its promise, components, processes, and problems. *IEEE Consumer Electronics Magazine, 7,* 6–14.
2. https://www.industryarc.com/Report/7385/industrial-internet-of-things-(IIoT)-market-report.html.
3. Arora, A., Kaur, A., Bhushan, B., & Saini, H. (2019). Security concerns and future trends of Internet of Things. In *2019 2nd International Conference on Intelligent Computing, Instrumentation and Control Technologies (ICICICT).* https://doi.org/10.1109/icicict46008.2019.8993222.
4. Lin, J., Yu, W., Zhang, N., Yang, X., Zhang, H., & Zhao, W. (2017). A survey on Internet of Things: Architecture, enabling technologies, security and privacy, and applications. *IEEE Internet of Things Journal, 4,* 1125–1142.

5. Miller, D. (2018). Blockchain and the Internet of Things in the industrial sector. *IT Professional, 20,* 15–18.
6. Huang, J., Kong, L., Chen, G., Wu, M., Liu, X., & Zeng, P. (2019). Towards secure industrial IoT: Blockchain system with credit-based consensus mechanism. *IEEE Transactions on Industrial Informatics, 15,* 3680–3689.
7. Xu, Y., Ren, J., Wang, G., Zhang, C., Yang, J., & Zhang, Y. (2019). A blockchain-based nonrepudiation network computing service scheme for INDUSTRIAL IoT. *IEEE Transactions on Industrial Informatics, 15,* 3632–3641.
8. Liang, W., Tang, M., Long, J., Peng, X., Xu, J., & Li, K. (2019). A secure fabric blockchain-based data transmission technique for industrial Internet-of-Things. *IEEE Transactions on Industrial Informatics, 15,* 3582–3592.
9. Yao, H., Mai, T., Wang, J., Ji, Z., Jiang, C., & Qian, Y. (2019). Resource trading in blockchain-based Industrial Internet of Things. *IEEE Transactions on Industrial Informatics, 15,* 3602–3609.
10. Zhao, S., Li, S., & Yao, Y. (2019). Blockchain enabled Industrial Internet of Things technology. *IEEE Transactions on Computational Social Systems, 6,* 1442–1453.
11. Al-Jaroodi, J., & Mohamed, N. (2019). Blockchain in industries: A survey. *IEEE Access, 7,* 36500–36515.
12. Fernández-Caramés, T. M., & Fraga-Lamas, P. (2019). A review on the application of blockchain to the next generation of cybersecure Industry 4.0 smart factories. *IEEE Access, 7,* 45201–45218.
13. Goel, A. K., Rose, A., Gaur, J., & Bhushan, B. (2019). Attacks, countermeasures and security paradigms in IoT. In *2019 2nd International Conference on Intelligent Computing, Instrumentation and Control Technologies (ICICICT)*. https://doi.org/10.1109/icicict46008.2019.8993338.
14. Shrouf, F., Meré, J.B., & Miragliotta, G. (2014). Smart factories in Industry 4.0: A review of the concept and of energy management approached in production based on the Internet of Things paradigm. In *2014 IEEE International Conference on Industrial Engineering and Engineering Management* (pp. 697–701).
15. https://www.iot-now.com/2018/10/12/89230-iot-iiot-connected-industry-industry-4-0-come-together-create-new-model-business/.
16. Preuveneers, D., & Zudor, E. I. (2017). The intelligent industry of the future: A survey on emerging trends, research challenges and opportunities in Industry 4.0. *Journal of Ambient Inteligence and Smart Environments, 9,* 287–298.
17. Li, Z., Kang, J., Yu, R., Ye, D., Deng, Q., & Zhang, Y. (2018). Consortium Blockchain for secure energy trading in Industrial Internet of Things. *IEEE Transactions on Industrial Informatics, 14,* 3690–3700.
18. Kang, J., Yu, R., Huang, X., Maharjan, S., Zhang, Y., & Hossain, E. (2017). Enabling localized peer-to-peer electricity trading among plug-in hybrid electric vehicles using consortium BLOCKCHAINS. *IEEE Transactions on Industrial Informatics, 13,* 3154–3164.
19. Bhushan, B., & Sahoo, G. (2020). Requirements, protocols, and security challenges in wireless sensor networks: An industrial perspective. In *Handbook of computer networks and cyber security* (pp. 683–713). https://doi.org/10.1007/978-3-030-22277-2_27.
20. Yu, Y., Chen, R., Li, H., Li, Y., & Tian, A. (2019). Toward data security in edge intelligent IIoT. *IEEE Network, 33,* 20–26.
21. Xu, L., He, W., & Li, S. (2014). Internet of Things in industries: A survey. *IEEE Transactions on Industrial Informatics, 10,* 2233–2243.
22. Müller, J. M., Kiel, D., & Voigt, K. (2018). What drives the implementation of Industry 4.0? The role of opportunities and challenges in the context of sustainability. *Sustainability, 10,* 247.
23. Kiel, D., Arnold, C., & Voigt, K. (2017). The influence of the Industrial Internet of Things on business models of established manufacturing companies—A business level perspective. *Technovation, 68,* 4–19.
24. Zelbst, P. J., Green, K. W., Sower, V. E., & Bond, P. (2019). The impact of RFID, IIoT, and Blockchain technologies on supply chain transparency. *Journal of Manufacturing Technology Management.*

25. Sun, Y., Zhang, L., Feng, G., Yang, B., Cao, B., & Imran, M. (2019). Blockchain-enabled wireless Internet of Things: Performance analysis and optimal communication node deployment. *IEEE Internet of Things Journal, 6,* 5791–5802.
26. Nash, P. (2017). Challenges of the Industrial Internet of Things. [Online]. Available: https://www.invma.co.uk/blog/iiot-challenges.
27. Isaja, M., & Soldatos, J. (2018). Distributed ledger architecture for automation, analytics and simulation in industrial environments. *IFAC-PapersOnLine, 51,* 370–375.
28. Gottheil, A. (2018). Can blockchain address the Industrial IOT security? [Online]. Available: https://iiot-world.com/cybersecurity/canblockchain-address-the-industrial-IOT-security.
29. Wang, F., Yuan, Y., Zhang, J., Qin, R., & Smith, M. H. (2018). Blockchainized internet of minds: A new opportunity for cyber-physical-social systems. *IEEE Transactions on Computational Social Systems, 5,* 897–906.
30. https://www.engineering.com/IOT/ArticleID/14023/IoT-Trends-Manufacturing-andAutomation-to-Lead-IoT-Growth-into-2017-and-Beyond.aspx.
31. Proof of Stake. (2018). An introduction to consensus algorithms: Proof of stake and proof of work. [Online]. Available: https://cryptocurrencyhub.io/an-introduction-to-consensus-algorithmsproof-of-stake.
32. Pasquali, E. (2018). Industrial IOT and the (data) sharing economy. [Online]. Available: https://iiot-world.com/connected-industry/industrialiot-and-the-data-sharing-economy.
33. https://rocketbug.com/google_business_view.
34. Urquhart, L., & McAuley, D. (2018). Avoiding the internet of insecure industrial things. *ArXiv, abs/1801.07207.*
35. Brass, I., Tanczer, L., Carr, M., Elsden, M., & Blackstock, J. J. (2018). Standardising a moving target: The development and evolution of IoT security standards. *IoT 2018.*
36. Varshney, T., Sharma, N., Kaushik, I., & Bhushan, B. (2019). Authentication & encryption based security services in blockchain technology. In: *2019 International Conference on Computing, Communication, and Intelligent Systems (ICCCIS).* https://doi.org/10.1109/icccis48478.2019.8974500.
37. Bassi, L. (2017). Industry 4.0: Hope, hype or revolution? In: *2017 IEEE 3rd International Forum on Research and Technologies for Society and Industry (RTSI)* (p. 1–6).
38. Zhou, L., Wu, D., Chen, J., & Dong, Z. (2018). When computation hugs intelligence: Content-aware data processing for industrial IoT. *IEEE Internet of Things Journal, 5,* 1657–1666.
39. Domova, V., & Dagnino, A. (2017). Towards intelligent alarm management in the age of IIoT. *Global Internet of Things Summit (GIoTS), 2017,* 1–5.
40. Brusakova, I. A., Borisov, A. D., Gusko, G. R., Nekrasov, D. Y., & Malenkova, K. E. (2017). Prospects for the development of IIOT technology in Russia. *IEEE Conference of Russian Young Researchers in Electrical and Electronic Engineering (EIConRus), 2017,* 1315–1317.
41. Zhou, L., & Guo, H. (2018). Anomaly detection methods for IIoT networks. In: *2018 IEEE International Conference on Service Operations and Logistics, and Informatics (SOLI)* (pp. 214–219).
42. https://www.scielo.br/scielo.php?script=sci_arttext&pid=S0103-65132018000100401.
43. Sajid, A., Abbas, H., & Saleem, K. (2016). Cloud-assisted IoT-based SCADA systems security: A review of the state of the art and future challenges. *IEEE Access, 4,* 1375–1384.
44. Quarta, D., Pogliani, M., Polino, M., Maggi, F., Zanchettin, A. M., & Zanero, S. (2017). An experimental security analysis of an industrial robot Controller. *IEEE Symposium on Security and Privacy (SP), 2017,* 268–286.
45. Eden, P., Blyth, A., Jones, K., Soulsby, H., Burnap, P., Cherdantseva, Y., & Stoddart, K. (2017). SCADA system forensic analysis within IIoT.
46. Lojka, T., Miskuf, M., & Zolotová, I. (2016). Industrial IoT gateway with machine learning for smart manufacturing. In: *APMS.*
47. Guan, Z., Lu, X., Wang, N., Wu, J., Du, X., & Guizani, M. (2019). Towards secure and efficient energy trading in IIoT-enabled energy internet: A blockchain approach. *Future Generation Computer Systems.*

48. https://www.forbes.com/sites/louiscolumbus/2016/12/03/industrial-analytics-based-on-internet-of-things-will-revolutionize-manufacturing/#20976d726c03.
49. Uhlmann, E., Hohwieler, E., &Geisert, C. (2017). Intelligent production systems in the era of Industrie 4.0—Changing mindsets and business models. *Journal of Machine Engineering*.
50. Banerjee, M., Lee, J., & Choo, K. R. (2017). A blockchain future for internet of things security: A position paper. *Digital Communications and Networks, 4,* 149–160.
51. Hossain, M. S., & Ghulam, M. (2016). Cloud-assisted Industrial Internet of Things (IIoT)—Enabled framework for health monitoring. *Computer Networks, 101,* 192–202.
52. Varshney, T., Sharma, N., Kaushik, I., & Bhushan, B. (2019). Architectural model of security threats & their countermeasures in IoT. In *2019 International Conference on Computing, Communication, and Intelligent Systems (ICCCIS)*. https://doi.org/10.1109/icccis48478.2019.8974544.
53. Sharma, T., Satija, S., & Bhushan, B. (2019). Unifying blockchian and IoT: Security requirements, challenges, applications and future trends. In *2019 International Conference on Computing, Communication, and Intelligent Systems (ICCCIS)*. https://doi.org/10.1109/icccis48478.2019.8974552.
54. Zhou, K., Liu, T., & Zhou, L. (2015). Industry 4.0: Towards future industrial opportunities and challenges. In *2015 12th International Conference on Fuzzy Systems and Knowledge Discovery (FSKD)* (pp. 2147–2152).
55. Chen, B., Wan, J., Shu, L., Li, P., Mukherjee, M., & Yin, B. (2018). Smart factory of Industry 4.0: Key technologies, application case, and challenges. *IEEE Access, 6,* 6505–6519.
56. Arora, D., Gautham, S., Gupta, H., & Bhushan, B. (2019). Blockchain-based security solutions to preserve data privacy and integrity. In *2019 International Conference on Computing, Communication, and Intelligent Systems (ICCCIS)*. https://doi.org/10.1109/icccis48478.2019.8974503.
57. https://www.statista.com/statistics/647231/worldwide-blockchain-technology-market-size/.
58. Yan, J., Meng, Y., Lu, L., & Li, L. (2017). Industrial big data in an Industry 4.0 environment: Challenges, schemes, and applications for predictive maintenance. *IEEE Access, 5,* 23484–23491.
59. Park, H., Kim, H., Joo, H., & Song, J. (2016). Recent advancements in the Internet-of-Things related standards: A oneM2M perspective ☆. *ICT Express, 2,* 126–129.
60. Jadon, S., Choudhary, A., Saini, H., Dua, U., Sharma, N., & Kaushik, I. (2020). Comfy smart home using IoT. *SSRN Electronic Journal*. https://doi.org/10.2139/ssrn.3565908
61. Kim, J., Yun, J., Choi, S., Seed, D., Lu, G., Bauer, M., et al. (2016). Standard-based IoT platforms interworking: Implementation, experiences, and lessons learned. *IEEE Communications Magazine, 54,* 48–54.

Recent Emerging Technologies for Intelligent Learning and Analytics in Big Data

Korhan Cengiz, Rohit Sharma, Kottilingam Kottursamy, Krishna Kant Singh, Tuna Topac, and Basak Ozyurt

Abstract During the past era of information technologies, the use of Big Data applications has increased in importance, as well as its phenomenon. Putting applications related Big Data to the force does not make it worth throwing this phenomenon into the background. But a great deal of work has been done to tackle the sophisticated challenges arising from the large quantities of data. Different types of software distributions and related technologies have been adopted after all. This chapter is an overview of recent emerging technologies for intelligent learning and analytics in multimedia and at the same time for online streaming processing of multimedia data for applications and system layers in urban life, particularly in education. These technologies are designed to help selection and usage of the right blend of multiple technologies for large volumes of data according to theirs needs and the requirements of concrete apps.

Keywords Big Data · Industry 4.0 · IoT · IoE

K. Cengiz (✉)
Faculty of Engineering, Trakya University, Edirne, Turkey
e-mail: korhancengiz@trakya.edu.tr

R. Sharma
SRM Institute of Science and Technology, SRM University, Ghaziabad, India

K. Kottursamy
SRM Institute of Science and Technology, Kattankulathur Campus, Chennai, India

K. K. Singh
KIET Group of Institutions, Ghaziabad, India

T. Topac
Department of Computational Sciences, Trakya University, Edirne, Turkey

B. Ozyurt
Faculty of Applied Sciences, Trakya University, Edirne, Turkey

R. Kumar et al. (eds.), *Multimedia Technologies in the Internet of Things Environment*, Studies in Big Data 79, https://doi.org/10.1007/978-981-15-7965-3_5

1 Introduction

Big Data continues to be an important research and innovation theme in several lines of business, chiefly because of the impact of recent innovations telecommunication technologies [1]. Result of this poorly understood density of data that can be captured from various sources, typically triggered by an increase in the volume of data collected in high gear, it is vital to have reliable computational methods for data mining operations. Spurred on by the tremendous acceleration of technology, the proliferation of large-scale data tools and frameworks, many discussions are taking place about high-performance large data request tools, especially those that are more appropriate for specific analytical purposes. In a few years, the mass of data in circulation has grown dramatically as a direct result of advances in ICT and the sharp uptake of the Internet, casual online operations, and multi-sourced operations [2]. In that regard, organizations are beginning to grasp how urgent it is to take leverage of evidence to support your data decision making in this domain [3]. Notwithstanding, the complexity and many associated challenges of the large-scale data field, it is virtually impossible to dismiss the available potential there [4]. Big Data for real world scenarios can increase the ability to acquire, hold, understand, process, and analyze huge amounts of data, and reveal hidden patterns that can be very effective in defining corporate strategies of analog universe. Big Data is major trend in information technology (IT) in earlier years, with a lot of labor input in relation to query and processing tools [5]. For example, as this has grown, the SQL computation, notably SQL-on-Hadoop, is extensively investigated and analyzed [6]. Different study has also separately been undertaken to assess the efficiency of some major data processing operations utilities. In that respect, in light of the Transaction Processing Council Decision(TPC-DS) [7], evaluate Impala [8], Hive [9], and Spark [10] to support comparison queries, comparing the fulfillment time for users. The outcomes of the project demonstrate that Impala is truly the sole engine that allowed for a truly interactive demand response in neither of the user cases. Hive based on MapReduce [5], SQL (Spark) based on a fault-tolerant distributed data set (trove of elements that can be accessed via the parallel nodes in cluster that operates), associated with Impala, which is distributed polling engine, beside the point verifying that they can give a quick answer to queries in Big Data mining. The use of scalable, highly distributed systems for the preparation of mass volumes of such data is recognized as one of the latest key trends in technology. Apache Hadoop MapReduce becomes the benchmark in this area, but it has disadvantages that are slowness and lots of requirements. It promotes contributions of relevant utilities for analysis through Hadoop. While some users use native query languages or application program interfaces (APIs), those systems recognize SQL as redeemer. SQL utilities are not same, which poses a considerable challenge in deciding which tool is most appropriate for a specific data analysis project. Figure 1 describes the 5V Concept.

For purposes of comparison and due to their popularity and use cases, some concepts behind Big Data such as Camel [11], Drill [12], Hive, Ignite [13], Impala

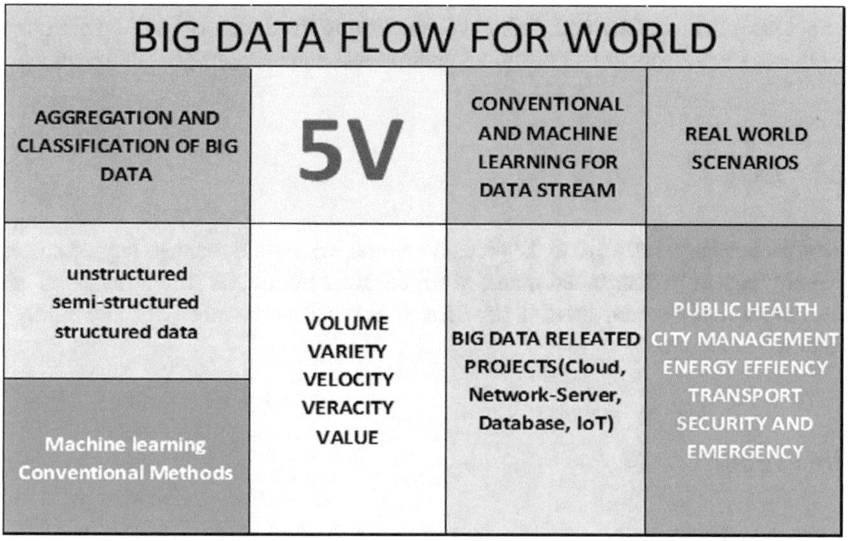

BIG DATA FLOW FOR WORLD			
AGGREGATION AND CLASSIFICATION OF BIG DATA	5V	CONVENTIONAL AND MACHINE LEARNING FOR DATA STREAM	REAL WORLD SCENARIOS
unstructured semi-structured structured data	VOLUME VARIETY VELOCITY VERACITY VALUE	BIG DATA RELEATED PROJECTS(Cloud, Network-Server, Database, IoT)	PUBLIC HEALTH CITY MANAGEMENT ENERGY EFFIENCY TRANSPORT SECURITY AND EMERGENCY
Machine learning Conventional Methods			

Fig. 1 Big Data flow world

[8], Kerby [14], Maven [15], Petri [16], PLC4X [17], Presto [18], and Spark are investigated in this chapter.

2 Promising Concepts for Big Data

2.1 Camel

Camel is a feature-rich open-source integration framework derived from well-established models of business integration. It lets users specify routing statue in a capacious menu of specific languages, including a Java-based APIs, XML configuration files. That infers that users get intelligent integration of routing rules in your integrated development environment (IDE), regardless of whether you use a both languages.

2.2 Drill

Common query engines need substantial IT interaction before data can be queried. With Drill, all this overwhelming workload is eliminated, so consumer can simply query the raw one on the field. Data is not needed to be loaded manually and schemas do not have to be maintained. Behalf of simplifying the directory Hadoop directory, databases that are document designed or simple storage service [19] in the query.

2.3 Hive

Hive data storage software makes it easy to read, write, and manage large data sets that are located in distributed memory with SQL. After all, its structure can be laid out onto already stored data. It provides Java database connectivity and surely a command line tool.

2.4 Ignite

Ignite is a horizontally scalable, fault-tolerant, high-performance distributed in-memory computing platform for creating real-time applications that can handle terabytes of data at an in-memory speed. With it as a high-performance distributed storage layer, you can accelerate existing services, applications, and APIs up to 100 times faster. You can synchronize the underlying databases by using Ignite as an in-memory data grid.

2.5 Impala

Impala sets the bar further for SQL query performance on Hadoop without sacrificing the ease of use you are accustomed to. Whether stored in Hadoop Distributed File System or HBase [20], Impala allows you to query data in real time—including SELECT, JOIN, and aggregate functions. In addition, it uses the same metadata, SQL syntax, open database connectivity driver, and user interface as Hive, providing a common and familiar platform for batch or real-time queries. (This allows minimal deployment.)

2.6 Kerby

Kerby is a Java Kerberos integration. It offers a comprehensive, highly interactive, and intuitive implementation, library, key distribution center (KDC) and various facilities that integrate public key infrastructure (PKI), one-time password (OTP), and token (OAuth2) as requested in any modern environment such as cloud, Hadoop, and mobile.

2.7 Maven

Maven is an indirect Big Data tool during working up with projects. It allows us to set a standard within the project, simplify the development process, create our documentation effectively, eliminate library dependency and IDE dependency. It gives you the flexibility to start a new project or module in seconds as it can be used across multiple projects, and there is no hassle for newbies in a project. It manages dependencies such as self-acting updates, closures (also known as transitive ones) and allows you work contently and organizedly. Projects that use Maven are growing, and its repository of libraries and metadata can be used as ground breaking. Agreements with the largest open-source projects for real-time availability of their latest releases are known.

2.8 Petri

The Petri committee (as in "Petri dish"—where cultures thrive and flow) assists outside project community stakeholders interested in becoming an Apache project to learn how the Apache software foundation (ASF) works, their thoughts on the community, and what it takes to build a healthy community in the long run. Its function is to advisory existing external communities ("cultures") on "The Apache Way" by emphasizing community leadership, which also involves discussions on ASF policy. The mentoring and training are done on a mailing list.

2.9 PLC4X

PLC4X is a suite of libraries for communication with industrial programmable logic controllers (PLCs) that use a number of protocols but with a common API.

2.10 Presto

Presto is a spread-out free SQL query engine for dynamic, interactive queries across a wide range of data types from gigabytes to petabytes, originally planned from zero for interactive analysis. More recently, companies like Facebook use it.

2.11 Spark

Spark provides high performance for data processing, making use of the latest database availability group scheduler, physical execution environment, and so on. It features over 80 high-level operators that facilitate the creation of parallel applications. And you can use it with various specific programming language for data science and SQL shells interactively. It can be run on Hadoop, Mesos [21], Kubernetes [22], standalone, or in the cloud. It has access to various data sources.

3 Big Data Definition Compared to Past

The concept of Big Data describes extensive, increasing data sets that encompass a heterogeneous format: unstructured, semi-structured, structured data [23]. Big Data's nature is complex, requiring high-capacity advanced computer technology and algorithms. Therefore, traditional tools cannot be more capable of being more efficient for longer.

3.1 5V Concept

3.1.1 Volume

Since the data is getting more and more large, the existent system runs into big challenges when ever it is called for to cope with an unparalleled volume of data. Today, we have a massive deal of demand for the design of efficient and intelligent forms of learning. So, we should meet the future demands [24].

3.1.2 Variety

The other dimension of data is diversity which means variety that keeps large data fascinating and challenged [25]. This phenomenon makes the data that usually originates from multiple sources and is of varying nature.

3.1.3 Velocity

Speed truly counted for Big Data poses new third hurdle for IoT. We have to complete a task within a given time period for our work; else the processing results become worse, like education, health system, and stock market forecasts [26].

3.1.4 Veracity

In the history, environmental methods such as machine learning algorithms have usually been developed with relatively accurate data from known and quite limited sources, so that learning outcomes also shows tendency to be infallible; therefore, truthfulness was never a serious problem which gives cause for concern [27]. Yet with the naked eye of the problems for twenty-first century, the directness and confidence of source data turn into a major problem like smoke, because the data and the sources often have many different roots and data quality is not always auditable.

3.1.5 Value

Indeed, through the use of a variety of learning methods for the analysis of large data sets, the final purpose is to obtain precious information from huge data volumes for economic interests. Therefore, the value is also referred to as outstanding feature of large data [28]. Nevertheless, obtaining a asked worth from large data sets in a low value density traffic is not easy.

4 Aggregation and Classification of Big Data

Synchronizing external data sources is a further challenge, and distributed platforms (including apps, repositories [29], sensors, networks, etc.) connected the inner structure of a setup are a reality. Examining data for organizations is not enough. To gain valuable insights and knowledge, it is essential to move one step farther and combine data sources. Outside data could be information on the health system, educational, and safety conditions, data from patients, students, and feedback from criminals. This can, for example, be useful in maximizing the strength of the models used for predictive analytics.

Classification of unbalanced data sets is another challenge. This issue has attracted a lot of public attention in recent years. Truth be told, applications of the teal world can create different degree in its environment. The first matter is often underrepresented with an insignificancy of minority and positive degrees. The second matter is largeness of major and negative degrees. The identification of minority matter is crucial in several areas, such as medical diagnosis [27]. Figure 2 shows the some promising Big Data projects in the literature.

APF BIG DATA RELEATED PROJECTS		
NAME	SUB CATEGORY	USED PROGRAMMING LANGUAGE
Accumulo		10
Airavata	Cloud, Network-Server	10
Ambari		10, 6, 11
Apex		10
Avro	Library	1, 3, 5, 10, 9, 6, 8
Beam		10, 6
Bigtop		10
BookKeeper		10
Calcite	Hadoop, Sql, Geospatial	10
CarbonData		10, 12
CrouchDB	Database, http, network-client, network-server, cloud, content	11, 4, 3, 1
Crunch	Library	10, 12
Drill		10
Flink		10, 12
Flume		10
Flou		10
Flou Recipes		10
Flou YARN		10
Giraph		10
Helix	Cloud	10
Ignite	IoT	10, 5, 3, 2, 12, 7
Kibble		6
Knox		10
Kudu		3
Lens		10
MetaModel	Database, Library	10
OODT		10
Oozie		10, 11
ORC	Database, hadoop, Library	10, 3
Parquet		10
Phoenix	Database	10, 2
Predictionio		12
REEF		10, 5, 3
Samza		12
Sqoop		10
Storm		10
Tajo		10
Tez		10
Trafodion		3
VXQUERY	xml	10
Daffodil	Library, xml	12, 10
DataFu		10
Edgent	Library, mobile, network-client	10, 11
DirectMemory		10
Falcon		10
Hama		10
Zeppelin		10, 11, 12
C(1972):1, SQL(1974):2, C++(1979):3, ERLANG(1986):4, C#(1990):5, PYTHON(1990):6, ODBC(1992):7, RUBY(1993):8, PHP(1994):9, JAVA(1995):10, JAVASCRIPT(1995):11, JDBC(1997):12, SCALA(2001):13		

Fig. 2 Big Data related projects [30]

5 Big Data Analytics

Huge data provides great chances and transformative potential for a range of disciplines; on balance, however, it is also an unexpected challenge to handle such enormous and growing volume of data. A more sophisticated data analysis is essential to help understand the correlations between the characteristics and to examine the data. For example, data analysis enables a healthcare system to gather precious information and monitor the samples that have a positive or negative impact on an illness. In addition, others also require real-time analysis, such as navigation, social networks, and educational systems. Therefore, modern algorithms and powerful data mining methods are required to generate precise results, monitor changes in different areas, and predict further field observations. Today, data mining, statistical analysis, visualization, and machine learning are available for users. Hence, Big Data drove the system architectures, hardware plus the software thorough today for future. Nevertheless, Big Data still needs to make progress to improve its analytical perspective in order to major data challenges and data stream processing.

6 Conventional and Machine Learning for Data Stream

Up-to-date applications from the real world such as sensor networks, network traffic, payment card transactions, inventory management, and blog posts generate enormous amounts of data. Methods of data mining are important for discovering attractive models and gaining values hidden in such gigantic data sets and data streams. Conventional data mining methods lack efficiency, scalability, and precision when applied to such large datasets. These are association mining, clustering, and classifying. Due to volume, speed, and flow variability, storing and analyzing are not possible for long term. So, scholars are faced with the necessity to hunt up new styles to improve analytical skills to manage data entities in a very constrained while to produce using constrained vaults (e.g., memory) and accurate results in terms of real data quality. Moreover, the fluctuation of currents leads to unforeseen modifications (i.e., a different distribution of instances) in incoming data streams.

Machine learning focuses on the theory, performance and characteristics of examining systems with algorithms. It draws on clues from diversity of related areas like mathematics and engineering. It has encompassed almost every scientific field, which has had a major effect on real world because it can be implemented in lots of applications [31]. It can be employed recommendation engines and recognition systems, computer science and data mining, and autonomous control systems [32]. This affects the exactness of the classification. Therefore, for tendency assessment, many methods have been used. Classification and clustering are the most known.

Granular computing (GrC) is quite all the go to use for various Big Data applications. If you are occupied in doing intelligent analysis, pattern recognition, it demonstrates many benefits. GrC plays an active role when it comes to designing models making decision while guaranteeing willed output. In technical way, GrC represent a broad calculation oriented toward about granulates such as clusters and intervals. In this manner, an efficient calculation model can be generated for complicated Big Data applications like data mining, remote sensing.

Incremental learning and ensemble learning are learning dynamic strategies. These are ordinary methods in learning from big stream data with approach drift. Incremental and ensemble learnings are widely applied to data streams and large-scale data. Both discusses various difficulties such as data availability and scarce resources. They are aligned for many applications such as prediction of inventory trends and user profile creation. The use of incremental learning allows quicker classification or prediction times when new data is available. Many machine learning algorithms natively support incremental one, and other algorithms could be tailored to accomplish work. Some examples of incremental algorithms are decision trees, decision rules, neural networks, neural Gaussian radial-based function networks or the incremental support vector machine.

Today, deep learning is an incredibly effective field of research in machine learning and its advanced levels of it. It integrals lots of parameters predictive analytics applications such as computer vision. The capability of traditional machine learning methods and features algorithms to process natural data for small data sets. On the opposite, deep learning is more productive to solve data analysis and learning problems found in enormous data sets. In actual fact, it aids in the automatic extraction of complex scenario representations from enormous quantities of unattended and uncategorized raw data. Beyond that, deep learning is multiple-stage learning and that deduces different levels of complex data abstractions, so that it is capable of facilitating the analysis of large amounts of data process related tasks of machine learning (e.g., a self-givenness extractor which has transformed the raw data) into a due inmost representation or attribute vector from which the learning subsystem, often a classifier, could identify or to sort patterns in the input. The training is not an oversight phase for deep learning in particular, as well. The learning algorithms are very difficult to parallelize because of repetitious calculations. Therefore, touching and scalable algorithms that are parallel must be created to the improve the training step of the deep learning. High voluminous data sets pose is a major challenge for deep learning. System deals with a scores of real inputs, wide varieties of outputs, and very high attributes. Thus, related solutions must overcome complexity. We can say that such vast amounts of information do not give an opportunity to train such a comprehensive algorithm with a CPU and memory. Owing to the several springs of Big Data, analytical researchers face other issues such as data which contains absentaneous inscriptions and noisy labels. We know that the data is processed with extremely high velocity and ought to be treated as it is alive. Besides that, in high speed, the data is often not steady and its distribution changes over time. Figure 3 shows some recent machine learning solutions for Big Data applications.

7 Big Data for Educational Use Cases

Knowledge discovery in databases (KDD) [33] is noted for its powerful role in the detection of hidden information from large data [34]. It has several applications such as e-commerce, bioinformatics, and educational data mining (EDM). Mining is very effective in both pedagogic and scholastic sector, especially in the study of online behavior in learning. Figure 4 illustrates some studies for educational data mining.

Comparison of machine learning technologies			
Learning types	Data processing tasks	Distinction norm	Learning algorithms
Supervised learning	Classification/Regression/Estimation	Computational classifiers	Support vector machine
Supervised learning	Classification/Regression/Estimation	Statistical classifiers	Naïve Bayes
Supervised learning	Classification/Regression/Estimation	Statistical classifiers	Hidden Markov model
Supervised learning	Classification/Regression/Estimation	Statistical classifiers	Bayesian networks
Supervised learning	Classification/Regression/Estimation	Connectionist classifiers	Neural networks
Unsupervised learning	Clustering/Prediction	Parametric	K-means
Unsupervised learning	Clustering/Prediction	Parametric	Gaussian mixture model
Unsupervised learning	Clustering/Prediction	Nonparametric	Dirichlet process mixture model
Unsupervised learning	Clustering/Prediction	Nonparametric	X-means
Reinforcement learning	Decision-making	Model-free	Q-learning
Reinforcement learning	Decision-making	Model-free	R-learning
Reinforcement learning	Decision-making	Model-based	TD learning
Reinforcement learning	Decision-making	Model-based	Sarsa learning

Fig. 3 Machine learning technologies out of Big Data [4]

List of studies that focused on educational data mining			
Objective	Platform	Data Mining Task	Source of Publication
Mining patterns of events in student's teamwork data	TRAC system	Sequential Pattern	Conference
Using data mining for automated chat analysis to understand support inquiry learning processes	Online chat	Classification	Conference
Discovering student preferences in e-learning	E-learning	Prediction	Conference
Mining the student online assessment data streaming environment using data mining and text mining	E-learning	Classification, Clustering, Association Rule Analysis	Conference
Examinig students' online interaction in a live video	Live video streaming environment	Clustering	Journal
A complete understanding of disorientation problems in web-based learning	Web-based learning system	Clustering	Journal
A web-based intelligent report e-learning system using data mining techniques	E-learning	Classification	Journal
Mining student data to characterize similar behavior groups in unstructured collaboration spaces	Ars Digita Community System	Clustering	Conference
Clustering and sequential pattern mining of online collaborative learning data	TRAC system	Clustering, Sequential Pattern	Conference

Fig. 4 Educational data mining studies [35]

8 Conclusions

The enhancement of Big Data applications has become significant nowadays. Putting several software solutions and applications have been performed as a result. This chapter proposes a summative assessment of recent emerging for intelligent learning, Big Data analytics in numerous applications in urban life. Such technologies are tailored to ease the selection and use of the right balance of multiple technologies for large volumes of data, matching their technological needs and specific application mandates.

References

1. Al Nuaimi, E., Al Neyadi, H., Mohamed, N., & Al-Jaroodi, J. (2015). Applications of big data to smart cities. *Journal of Internet Services and Applications, 6*(1), 1–15. https://doi.org/10.1186/s13174-015-0041-5.
2. Berberidis, D., Kekatos, V., & Giannakis, G. B. (2016). Online censoring for large-scale regressions with application to streaming big data. *IEEE Transactions on Signal Processing, 64*(15), 3854–3867. https://doi.org/10.1109/TSP.2016.2546225.
3. Bholat, D. (2015). Big Data and central banks. *Big Data and Society, 2*(1). https://doi.org/10.1177/2053951715579469.
4. Qiu, J., Wu, Q., Ding, G., Xu, Y., & Feng, S. (2016). A survey of machine learning for big data processing. *EURASIP Journal on Advances in Signal Processing, 2016*(1). https://doi.org/10.1186/s13634-016-0355-x.
5. Hu, H., Wen, Y., Chua, T. S., & Li, X. (2014). Toward scalable systems for big data analytics: A technology tutorial. *IEEE Access, 2,* 652–687. https://doi.org/10.1109/ACCESS.2014.2332453.
6. Rodrigues, M., Santos, M. Y., & Bernardino, J. (2019). Big data processing tools: An experimental performance evaluation. *Wiley Interdisciplinary Reviews: Data Mining and Knowledge Discovery, 9*(2). https://doi.org/10.1002/widm.1297.
7. Shi, L., Huang, B., Xu, H., & Ye, X. J. (2010). Implementation of TPC-DS testing tool. In *2010 2nd International Workshop on Database Technology and Applications, DBTA2010—Proceedings*. https://doi.org/10.1109/DBTA.2010.5658943.
8. Impala. https://impala.apache.org/overview.html. Accessed June 09, 2020.
9. Apache Hive TM. https://hive.apache.org/. Accessed June 09, 2020.
10. Apache Spark™—Unified Analytics Engine for Big Data. https://spark.apache.org/. Accessed June 10, 2020.
11. Apache Camel User Manual—Apache Camel. (2020). https://camel.apache.org/manual/latest/. Accessed June 09, 2020.
12. Apache Drill—Schema-Free SQL for Hadoop, NoSQL and Cloud Storage. https://drill.apache.org/. Accessed June 09, 2020.
13. Open Source In-Memory Computing Platform—Apache Ignite®. https://ignite.apache.org/. Accessed June 09, 2020.
14. Welcome to Apache Kerby—Apache Directory. https://directory.apache.org/kerby/. Accessed June 10, 2020.
15. Maven—Maven Features. https://maven.apache.org/maven-features.html. Accessed June 09, 2020.
16. (No Title). https://petri.apache.org/. Accessed June 10, 2020.
17. PLC4X. https://plc4x.apache.org/. Accessed June 10, 2020.
18. Presto | Distributed SQL Query Engine for Big Data. https://prestodb.io/. Accessed June 09, 2020.
19. Cloud Object Storage | Store & Retrieve Data Anywhere | Amazon Simple Storage Service (S3). https://aws.amazon.com/s3/. Accessed June 10, 2020.
20. Apache HBase—Apache HBase™ Home. https://hbase.apache.org/. Accessed June 10, 2020.
21. Apache Mesos. https://mesos.apache.org/. Accessed June 10, 2020.
22. Production-Grade Container Orchestration—Kubernetes. https://kubernetes.io/. Accessed June 10, 2020.
23. Gandomi, A., & Haider, M. (2015). Beyond the hype: Big data concepts, methods, and analytics. *International Journal of Information Management, 35*(2), 137–144. https://doi.org/10.1016/j.ijinfomgt.2014.10.007.
24. The Rise of Data: Data Science, Big Data and Data Analytics for Seamless Business Operations. https://www.analyticsinsight.net/the-rise-of-data-data-science-big-data-and-data-analytics-for-seamless-business-operations/. Accessed June 11, 2020.

25. Council Post: Location Intelligence is Becoming Democratized—Here's How Your Business Can Harness It. https://www.forbes.com/sites/forbestechcouncil/2020/06/11/location-intelligence-is-becoming-democratized---heres-how-your-business-can-harness-it/#5bc8d8961754. Accessed June 11, 2020.
26. Ways To Grow Your Business Globally With Big Data. https://yourstory.com/mystory/grow-business-globally-big-data. Accessed June 11, 2020.
27. Utilizing the Big Data Card in the Time of Coronavirus Pandemic. https://www.analyticsinsight.net/utilizing-bigdata-card-time-coronavirus-pandemic/. Accessed June 11, 2020.
28. Four Trends Shaping the New Era of Big Code—eWEEK. https://www.eweek.com/development/four-trends-shaping-the-new-era-of-big-code. Accessed June 11, 2020.
29. The World's Leading Software Development Platform—GitHub. https://github.com/. Accessed June 9, 2020.
30. Apache Projects List. https://projects.apache.org/projects.html?category#big-data. Accessed June 2, 2020.
31. Rudin, C., & Wagstaff, K. L. (2014). Machine learning for science and society. *Machine Learning, 95*(1), 1–9. https://doi.org/10.1007/s10994-013-5425-9.
32. Bishop, C. M. (2006). *Pattern recognition and machine learning*. New York: Springer.
33. Kottursamy, K., Raja, G., Padmanabhan, J., & Srinivasan, V. (2017). An improved database synchronization mechanism for mobile data using software-defined networking control. *Computers & Electrical Engineering, 57,* 93–103.
34. Kottursamy, K., Raja, G., & Saranya, K. (2016). A data activity-based server-side cache replacement for mobile devices. In *Artificial intelligence and evolutionary computations in engineering systems* (pp. 579–589). New Delhi: Springer.
35. Mohamad, S. K., & Tasir, Z. (2013). Educational data mining: A review. *Procedia Social and Behavioral Sciences, 97,* 320–324. https://doi.org/10.1016/j.sbspro.2013.10.240.

Real-Time Health System (RTHS) Centered Internet of Things (IoT) in Healthcare Industry: Benefits, Use Cases and Advancements in 2020

G. Arun Sampaul Thomas and Y. Harold Robinson

Abstract The usage of the IoT in healthcare (the industry, personal healthcare and healthcare payment applications) has gruffly increased across innumerable precise Internet of Things use cases. At the same time, we see how other healthcare IoT use cases are picking up speed and the connected healthcare veracity is fast-tracking, even if obstacles remain. Thus far, most IoT enterprises in healthcare revolved around the development of care as such with remote monitoring and telemonitoring as main applications in the wider scope of telemedicine includes real-time health systems (RTHS). A second area where many enterprises exist is tracking, monitoring and maintenance of assets, using IoT and RFID. This is done on the level of medical devices and healthcare assets, the people level and the non-medical asset level (e.g., hospital building assets). Furthermore, we have stated a few drivers and benefits that influence the evolutions and forecasts in the implementation of IoT in healthcare. Most of them fit in overall healthcare drivers such as aging populations, the changing behavior and demands of patients (consumerization of healthcare and patient-centricity) and of healthcare workers, budgetary challenges and the improvement of care quality.

Keywords Internet of Things · Medical solutions · Healthcare · RFID · Telemedicine · RTHS

1 Introduction

By 2020, 95% of healthcare organizations will have adopted IoT technology [1]. However, these deployments and use cases are just the beginning and, at the same time, are far from omnipresent. More advanced and integrated approaches within

G. Arun Sampaul Thomas (✉)
CSE Department, J.B. Institute of Engineering and Technology, Hyderabad, India
e-mail: arunthomas.cse@jbiet.edu.in

Y. Harold Robinson
School of Information Technology and Technology, Vellore Institute of Technology, Vellore, India
e-mail: yhrobinphd@gmail.com

R. Kumar et al. (eds.), *Multimedia Technologies in the Internet of Things Environment*, Studies in Big Data 79, https://doi.org/10.1007/978-981-15-7965-3_6

the possibility of the digital transformation of healthcare are starting to be used with regards to health data aspects where IoT plays an increasing role, as it does in specific applications such as smart pills, smart home care, personal healthcare, robotics and real-time health systems (RTHS).

1.1 Toward a More Cohesive and Advanced IoT-Enabled eHealth Veracity

Within the overall connected healthcare and eHealth depiction, more integrated approaches and benefits are sought with a role for the so-called Internet of Healthcare Things (IoHT) or Internet of Medical Things (IoMT) as shown in Fig. 1. The period from 2017 until 2022 will be important in this transition, with several changes before 2020. From 2017 until 2022, growth in IoT healthcare applications is indeed poised to

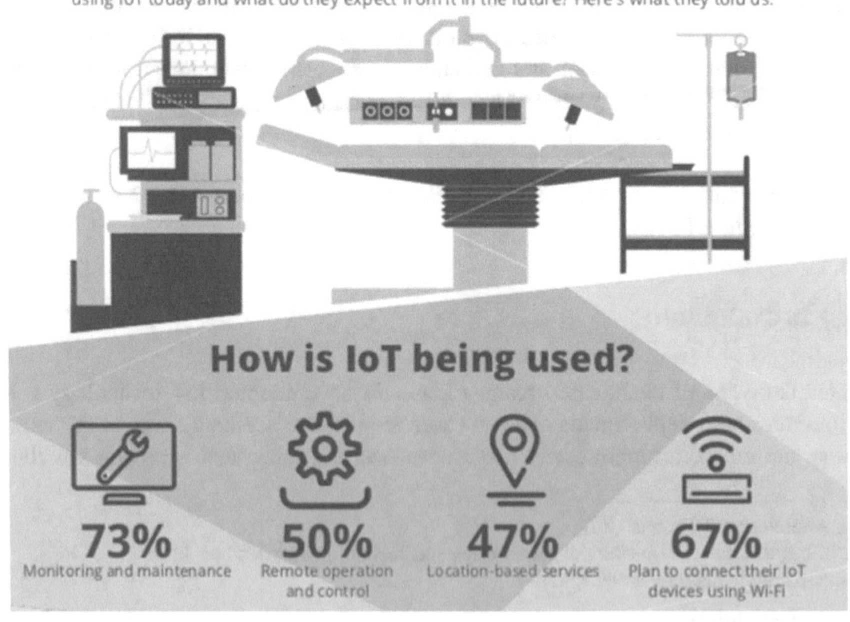

Fig. 1 IoT in the field of healthcare

accelerate as the Internet of Things is a key component in the [2] digital transformation of the healthcare industry and various stakeholders are stepping up their efforts. Moreover, there is an increasing consciousness and engagement of consumers with regards to their health, demand for remote and home possibilities keeps growing, various healthcare ecosystem players come up with novel approaches and partnerships, and healthcare expenditure reduction remains a main goal, along with better quality care. A more integrated and IoT-enabled eHealth approach proves essential in all these areas.

Some regions, such as the USA, are leading in healthcare IoT and in leveraging health-related data from IoT devices. According to the 2017 Thales Data Threat Report, healthcare edition, which we covered in an article on [3] healthcare and cybersecurity, already 30% of healthcare organizations use IoT for sensitive data.

And, as mentioned in our article, on how [4] healthcare providers and payers are boosting their digital transformation deployments in 2017, already by 2019 over 40% of healthcare organizations are expected to use IoT-enabled biosensors as IDC predicts.

March 2017 [5] research from Aruba Networks (more below) found that by 2019, 87% of healthcare organizations will have adopted Internet of Things technology and 76 percent believe it will transform the healthcare industry.

More data and evolutions in the mentioned articles. Let us now look at some key aspects and use cases within the broader scope of the Internet of Things in the healthcare sector in the following sections.

2 The Massive Landscape of Healthcare Stakeholders and IoT Prospects

We once started listing all the applications, examples and use cases regarding the usage of the Internet of Things (IoT) in healthcare. We had to stop. The reason: healthcare is such a vast ecosystem and once you also start including personal healthcare, the pharmaceutical industry, healthcare insurance, RTHS, healthcare building facilities, robotics, biosensors, smart beds, smart pills, anything remote and the various healthcare specializations, activities and even (treatments of) diseases, that list of Internet of Things applications in healthcare quickly becomes endless.

In the scope of this article, we do not focus on pharma nor on how IoT (and AI) helps in the treatment of specific diseases or the evolutions with regards to specific areas such as bionic limbs and others.

The mentioned vastness goes both for the IoT use cases and the real-life applications of the Internet of Things in the healthcare industry. Still, as said some use cases clearly stick out if we look at applications and evolutions on the side of healthcare providers and healthcare payers.

3 Hastening in All IoT Use Case and Applications in Healthcare Upfront

Two things are for sure: (1) the main IoT use case in healthcare for now is (remote) health monitoring, certainly from an [6] **IoT spending** perspective and (2) the Internet of Things will soon be pretty ubiquitous in healthcare and health-related activities and processes on various levels.

Moreover, we see that devices and IoT applications, which traditionally fit more in a consumer context, (e.g., personal health trackers) get an increasing place in the relationship between consumer/patient and healthcare providers and payers. Patient engagement and consumer consciousness play an important role here and in the relationship with healthcare payers also incentives and premiums do (compare with the use of telematics in insurance).

Outside of this scope, there is major growth ahead in a more [7] Industrial Internet of Things context, whereby healthcare providers, such as hospitals, leverage IoT, in combination with applications and technologies in the field of robotics, artificial intelligence and big data. The second focus area of IoT applications we mentioned in the introduction (monitoring, tracking, maintenance and so forth) is certainly also going to keep growing; albeit at difference paces, depending on the hospital, country and so on. Some will start with tracking anything from medical equipment and patients to hospital building assets and beds; others will move to the next stages.

4 IoT in the Framework of Healthcare Revolution and the Challenges of Information-Driven Healthcare

Remote health monitoring and various other IoT use cases in healthcare need to be seen in **the context of the main challenges in healthcare and the areas with a clear benefit and/or purpose/possibility of innovation**.
The developments regarding the Internet of Things in healthcare also need to be seen in the context of the digital transformation of the various healthcare segments. IoT, from an enablement perspective, is a cornerstone of the digital transformation of healthcare until at least the next decade.

The Internet of Things and healthcare information systems
In a health data context de facto quite, some data from medical devices and monitoring systems ultimately end up in electronic healthcare records (EHR) systems or in specific applications which are connected with them and send the data to laboratories, doctors, nurses and other parties involved.

As health-related data is collected and increasingly is available in real time, it gets integrated with electronic healthcare records (EHR).

EHR systems are far from omnipresent and most have not been designed with the Internet of Things, RFID and real-time data in mind; they have been designed, if all is well, to make healthcare faster, more patient-centric, more affordable and better from the perspective of the patient's health and the work of healthcare professionals, based on rather static data.

These outcomes are also essential in many IoT use cases in healthcare, yet they are not always achieved. Moreover, there are so many approaches to the digitization of healthcare records that in practice an Internet of Things deployment needs to take into account these differences if it is related with an individual patient.

Not all health data from connected devices ultimately lands in the EHR/EMR environment. There are plenty of other information systems and systems of insight, depending on type of data, device, scope and purpose. Moreover, there is a shift toward real-time health systems (RTHS), which go beyond EHR and include awareness and real-time data capabilities in an IoT and connected/wearable device perspective. EHR systems are part of the broader context and processes within this RTHS systems approach.

According to [8] Mind Commerce research, 'RTHS will be a key area for IoT in healthcare as Big Data Analytics tools and processes are utilized to evaluate both dynamic and static data for predictive analytics as part of comprehensive healthcare systems improvement programs.'

Healthcare data: working with purpose and security in mind

A second challenge has to do with the data. Healthcare information is very personal, and the selection of the required data needs to happen with the outcomes in mind.

It is the data that makes sense to improve the lives of patients and the organization of healthcare across its various aspects such as the ability for doctors, specialists, nurses and staff to make better decisions faster.

Moreover, security and privacy by design need to be part of any IoT use case, project or deployment. Leveraging the IoT and data aims to improve and reduce errors and costs. Making sure it does not get exposed or used for the wrong reasons is key.

As mentioned in other articles, personal healthcare data needs to be treated differently from a security and compliance perspective.

Special attention for personal data in healthcare IoT projects needed

Various regulations across the globe drive the compliance agenda, yet healthcare data security needs to go beyond compliance. At the same time, healthcare organizations need to pay more attention to compliance as well, certainly in regions where stricter regulations are being put into place such as the EU [9] General Data Protection Regulation (GDPR) where personal health data, as well as genetic and other medical and biological data, gets special attention and is seen as very sensitive.

It is clear that any IoT project which involves personal health data needs to take these rules and the lawfulness, intent, and diffusion stipulations, to name a few, into account. As a matter of fact, any IoT project should have security and privacy by design in mind where it concerns personal data. However, the positions with regards to the protection and leverage of health data are, to put it mildly, very different if you start comparing initiatives and regulations across the globe.

5 The IoT in Healthcare: Use Cases and Major Developments

Taking all the above into account let us take a look at the major use cases today and, next, the rapidly emerging use cases in the near future.

Remote health monitoring and telehealth
As mentioned, remote health monitoring for now is the major IoT use case in the Internet of Healthcare Things (IoHT). In other words: today's major use case from an IoT spending perspective is outside the setting of a hospital or other healthcare facility. There is a general shift of care in hospitals or emergency care environments to private environments such as the patient's home, whenever that becomes possible.

It is a matter of costs, it is a matter of getting the patient back to his 'normal environment' and it is one way to reduce the workload of healthcare workers who in many countries and many periods simply can't cope. In some countries, the lack of funding and, as a result, shortage of healthcare workers, specifically in and after seasons where more diseases strike, is a recurring yearly disaster. Remote health monitoring, which is obviously very possible thanks to the Internet of Things, also partially helps solve the rise of chronic diseases, among others due to an aging population (but not just that). Remote health monitoring is also ideal when patients live in remote areas. There is a broad range of (specialized) wearables and biosensors, along with other medical devices, available today that enables remote health monitoring.

Remote health monitoring also offers healthcare stakeholders the possibility to detect patterns, leveraging the data coming from these wearables and other devices. This enables new insights and visualizations of patterns as the combination of (big) data, analytics, IoT and so forth tends to do.

It is one reason why the skillsets of, for example, hospital staff, are changing. In practice, as budgets in healthcare are restricting, this could lead to concerns from a human care perspective.

The connected and smart hospital—a extensive variety of applications

Speaking about hospitals, there is a set of IoT use cases which brings us back from the remote aspect to hospitals and other [10] healthcare facilities.

From predictive maintenance of healthcare kit to smart beds

So, this is not one specific IoT use case [10] but rather a range which we could call the smart or connected hospital.

On the most essential level, which we touched upon earlier, we are speaking about RFID and IoT-enabled devices, IoT-enabled 'assets' and rather traditional general IoT use cases which are really cross-industry such as (predictive) maintenance of hospital assets, connected healthcare devices and the tracking of healthcare devices (and people).

It is not really a surprise that this is an ongoing evolution, which on other levels, also includes phenomena such as smart beds, the aggregation and real-time availability of data from healthcare devices and assets regarding specific patients, and the advent of robots in a hospital environment for routine tasks.

Applications in the asset- and information-intensive hospital

After all, the hospital is an asset- and device-intensive environment with medical equipment and a wide range of objects that can be connected and monitored in order to achieve tangible benefits.

In such asset-intensive environments which at the same time are extremely information-intensive, ample possibilities and potential outcomes emerge when leveraging IoT and related sets of technologies.

In a cross-industry perspective, we can certainly also mention smart buildings and facility management here. If there is one place where the various building parameters (think about temperature, humidity, air regulation, specific environmental controls, security and so forth) need to be optimal, it for sure is the hospital.

According to the previously mentioned 2017 [1] research by Aruba Networks on the state of IoT in healthcare (and more), the main IoT use case in healthcare organizations is monitoring and maintenance (73% of respondents), followed by remote operation and control (50% of respondents). Connecting IoT devices is indeed an important goal to reap the benefits from IoT with 67% of respondents planning to connect their IoT devices using Wi-Fi.

If we look at the most common IoT technologies [10], we see patient monitors (64%), energy meters (56%) and X-rays and imaging devices (33%).

The major perceived benefits, finally, are increased innovation (80%), visibility across the organization (76%) and, as always but in healthcare organization even more important, cost savings (Portrayed in Fig. 2).

The Internet of Healthcare Things and the twin function of wearables

Imagine that all medical equipment and devices are connected (within an RTHS environment) in a way that makes sense for the desired outcomes. Imagine the possibilities of a true Internet of Healthcare Things (some people speak about the Internet of Medical Things or IoMT to describe this reality of the connection of health-related devices and equipment, others call it the medical Internet of Things or mIoT).

Regardless of how you call it, it is clear that connecting all devices, apps, data and so forth opens up a new world of opportunities, as the Internet of Things in general does. The Internet of Healthcare Things includes the devices and connected assets (and their use cases) which we mentioned before such as everything that is needed for remote health monitoring, X-rays and imaging devices, the list goes on. In a sense, the Internet of Medical Things is already here—and has been for some time. However, if you really want to look at it from a holistic perspective, it is far from here and ample challenges need to be tackled: health device interoperability, integrated systems connecting patients and healthcare workers, secure standards, RTHS integration and so on.

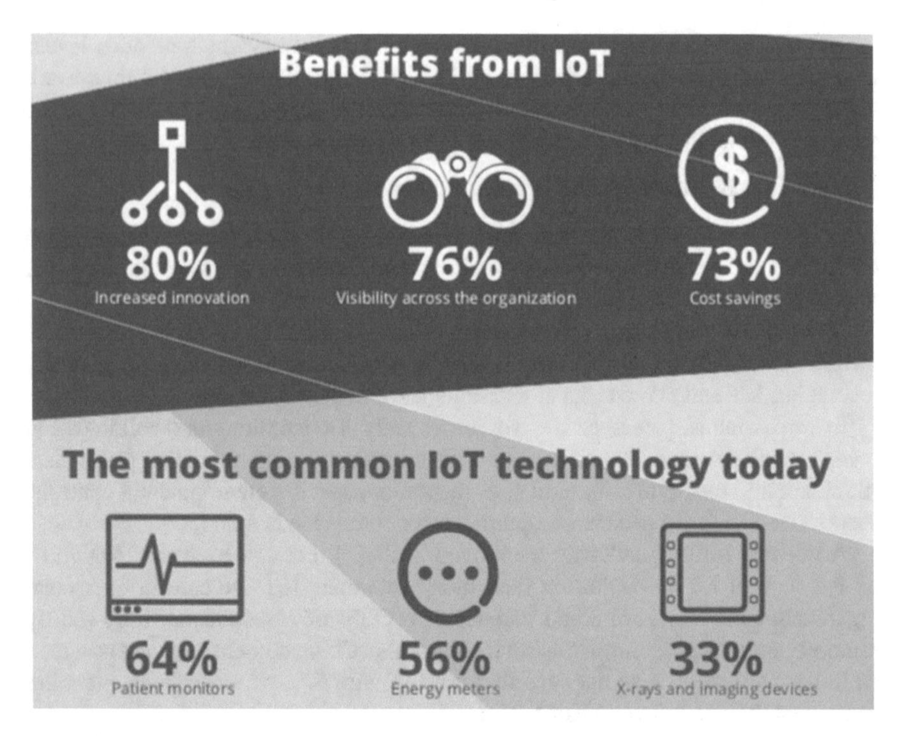

Fig. 2 Benefits of IoT and its common technology

What does IoHT encompass? As the name indicates virtually all healthcare things. Of course, we need to go beyond the healthcare or medical 'things' aspect and look at the broader picture. However, if we look at it on a device level, we see how the personal and 'consumer' IoT health sphere (remember the consumerization of healthcare in this sense and the increasing attention for devices, mainly wearables, which enable personal health monitoring and meet increasing 'consciousness' demands) and the health sector sphere where wearables, several biosensors and so forth are also increasingly important (professional ones). In that meeting of both spheres, new healthcare provider and healthcare payer models arise, including rather consumer-oriented wearables and those more professional devices.

On a wearable level, we also see an increasing use of specialized wearables across all parts of the body (next generation hearing aids, implantable wearables, skin patches, smart contact lenses, etc.).

Internet of Things meets robotics in healthcare

If we start looking at the (near) future, we see new healthcare use cases where IoT and robotics meet. This is already the case on some levels, but research indicates we are at the verge of far more levels.

By 2020, there will be 50% increase in the use of Robots to delivered medications, supplies and food throughout the hospital.

Among the many robots that are already in use today, there are well-known examples of medical robots in surgery (precision surgery or distance surgery), robots that are used for rehabilitation and hospital robots such as Panasonic's HOSPI that, among others, takes care of deliveries (medication, drinks, etc.). This latter category is poised for growth. According to IDC's Worldwide Healthcare IT 2017 Predictions, by 2019, there will be a 50% increase in the use of such robots who carry out tasks such as medication delivery, food delivery and delivery of supplies overall. In other words: taking care of rather routine tasks, freeing up (human) resources. Note: with routine we do not mean that these tasks are not important. For many hospitalized people, getting their medication or food delivered is a moment of human interaction they do value.

The global smart healthcare market by 2020—source and more information.

6 Conclusion and Imminent IoT Healthcare in 2020

We have mentioned a few drivers and benefits that influence the evolutions and predictions in the deployment of IoT in healthcare.

Most of them fit in overall healthcare drivers such as aging populations, the changing behavior and demands of patients (consumerization of healthcare and patient-centricity) and of healthcare workers, budgetary challenges and the improvement of care quality.

The previously mentioned research from Aruba Networks found that the large majority of perceived IoT benefits in healthcare today revolves around increased innovation, visibility across the organization and cost savings. On top of those, future IoT benefits which are expected include higher workforce productivity, the creation of new 'business models' and better collaboration.

End 2016, Technavio looked at the global smart healthcare market and identified four factors:

1. Higher demand for remote health monitoring of an aging population.
2. Increased consumer health consciousness.
3. Growing popularity of healthcare wearables.
4. Strategic alliances in the smart healthcare market, where the creation of those new business models also comes in.

As the graphic from the press release illustrates, increased use of remote patient monitoring will help drive the global market at a Compound Annual Growth Rate (CAGR) of 24.55% by 2020. Also, the advent of smarter technologies plays a role. The smart pills segment alone is expected to reach $6.93 billion by 2020 with a CAGR of 23.62%.

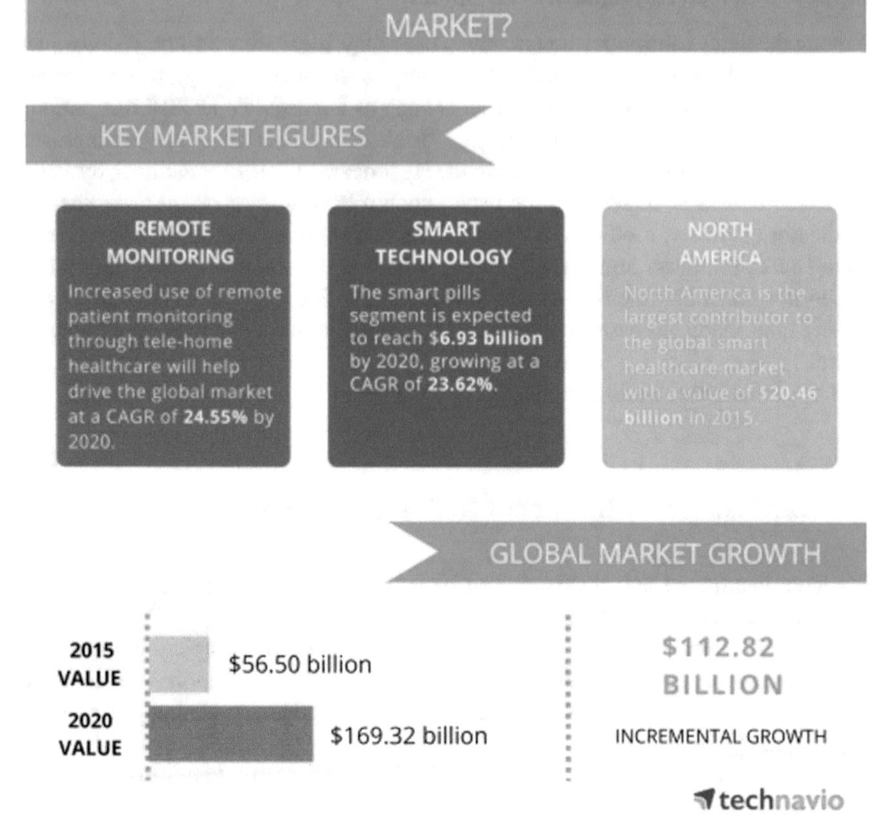

Fig. 3 Global market healthcare value. *Source* Technavio Research

Overall smart healthcare market value by 2020 is estimated to be $169.32 billion by 2020. A major part of it will be for remote patient monitoring as depicted in Fig. 3.

References

1. Zdravkovic, M., Noran, O., & Trajanovic, M. (2014). On pervasive health information systems in the internet of things. In *Proceedings of the 25th Australasian Conference on Information Systems* (pp. 1–10), December 8–10, 2014, Auckland University of Technology, Auckland, New Zealand.
2. Harold Robinson, Y., Santhana Krishnan, R., Golden Julie, E., Kumar, R., Son, L. H., & Thong, P. H., Neighbor knowledge-based rebroadcast algorithm for minimizing the routing overhead in mobile ad-hoc networks. *Ad Hoc Networks.*/newpage
3. Harold Robinson, Y., & Golden Julie, E. (2019). MTPKM: Multipart trust based public key management technique to reduce security vulnerability in mobile ad-hoc networks. *Wireless*

Personal Communications, 1–22.

4. Richardson, J. E., Ves, J. R., Green, C. M., Kem, L. M., Kaushal, R., & HITEC Investigators. (2015). A needs assessment of health information technology for improving care coordination in three leading patient-centered medical homes. *Journal of the American Medical Informatics Association, 22*, 815–820.
5. Mishuris, R. G., Yoder, J., Wilson, D., & Mann, D. (2016). Integrating data from an online diabetes prevention program into an electronic health record and clinical workflow, a design phase usability study. *BMC Medical Informatics and Decision Making, 16,* 1–13.
6. Ayyanar, A., Archana, M., Harold Robinson, Y., Golden Julie, E., Kumar, R., & Son, L. H. (2019). Design a prototype for automated patient diagnosis in wireless sensor networks. *Medical & Biological Engineering & Computing.*
7. Harold Robinson, Y., Golden Julie, E., Kumar, R., & Son, L. H. (2019). Probability-based cluster head selection and fuzzy multipath routing for prolonging lifetime of wireless sensor networks. *Peer-to-Peer Networking and Applications*, 1–15.
8. Corrin, L., Kennedy, G., & de Barba, P. (2017). Asking the right questions of big data in education. https://pursuit.unimelb.edu.au/articles/asking-the-right-questions-of-big-data-in-education.
9. Marr, B. (2015). How big data is changing healthcare. *Forbes/Tech.* https://www.forbes.com/sites/bernardmarr/2015/04/21/how-big-data-is-changing-healthcare/#6643a6972873.
10. Arun Sampaul Thomas, G., & Harold Robinson, Y. (2020). IoT, big data, blockchain and machine learning besides its transmutation with modern technological applications. In *Springer book chapter-internet of things and big data applications. Part of the Intelligent Systems Reference Library Book Series* (ISRL, Vol. 180, pp. 47–63), February 25, 2020. ISSN: 978-3-030-39118-8.

Building Intelligent Integrated Development Environment for IoT in the Context of Statistical Modeling for Software Source Code

Raghavendra Rao Althar and Debabrata Samanta

Abstract With various challenging areas as the focus in software development, statistical modelling for software source code tries to derive the knowledge hidden in various software artifacts including code and text and help building robust systems which are intelligent. Integrated Development Environment (IDE) is one area of focus in the domain. Since this provides platform for entire software development processes, it plays key role. In this chapter, study of intelligent IDEs is extended to IoT environment context. To begin with, we focus our study on understanding the IDEs and their capability. Exploration will focus on open-source IDE for mobile, and we look for better understanding the landscape so that we can extend the learning to the IoT world. In this work, building low cost IDE for mobile is focused open. Also, interestingly, exploration focuses on open-source components and possibility of putting them together. Then, in next part, we explore the IDE for Internet of things (IoT), with focus on open source ecosystem. In this review, we extend the exploration of IoT in to device management, data management, communication, intelligent data processing, security and privacy, and application deployment areas. This gives a greater insight in to the IoT world to extend the need of intelligent IDEs to IoT world. Then, to get the context of IDE in the machine learning context, we explore the topic of building optimal IDE for feature engineering which is one of the key phases in machine learning life cycle. Since machine learning projects are highly data-oriented eco system, learning of IDE and its insights in this area will provide rich insights in to the main theme of discussion which is intelligent IDE for IoT eco system.

Keywords Statistical modelling · Round time live · Internet of things · Weka · Multi-armed bandit · Procedural graphics · Service-oriented architecture

R. R. Althar
QMS, First American India, Bangalore, India
e-mail: ralthar@firstam.com

D. Samanta (✉)
Department of Computer Science, CHRIST (Deemed to be University), Bangalore, India
e-mail: debabrata.samanta369@gmail.com

R. Kumar et al. (eds.), *Multimedia Technologies in the Internet of Things Environment*, Studies in Big Data 79, https://doi.org/10.1007/978-981-15-7965-3_7

1 Introduction

Statistical modeling for software source code work is targeting to address all the issues faced by the software development processes. Primary areas of the application are focusing on addressing the challenge of low confidence that the community is facing today on the quality of the deliverables across the life cycle of software development. To build in transparency and confidence using the knowledge hidden in the software code is the focus. Lack of a better understanding of the business needs contributes to the challenge. Lack in the domain and technical knowledge for needed analysis contributes toward disconnect in understanding customer needs. This points toward maintaining the traceability of the software requirements across the life cycle. A big picture of the final delivery is missing, and the focus is more on the structure of the code rather than the holistic design approach. The gap between real-world user needs and the technical descriptions of the system is evident. Business domain understanding becomes key here. There is an effort toward making software development a community experience a customer point of view. Though there is an effort toward the same, the results are not satisfying. Business System Analyst who is supposed to take care of this, but due to lack of robust end-to-end system, this has not been practical. This hints at building an intelligent system that is traceable end to end and helps integrate all the knowledge across and feed in to, as needed to make meaningful decisions. Uncommon focus on unit testing is another important part of building a robust construction phase. Though the practice exist, they are not robust, for the case of Continuous Integration and Continuous Deployment (CICD), quality assurance focus takes back seat. Defects are handled as new requirements, which make people lose visibility on the quality progress of the work. Maturity of the CICD practice is a concern leading to the question of, the skill of the people during the transition from traditional methods to CICD way. First Time Right is focused but the true sense is missing in the operations, resulting in sub-optimal deliverables across phases of the life cycle. The impact of the lack in initial analysis extends itself into non-functional aspects of the requirements leading to security vulnerabilities costing organizations on reputation and money. Lack of knowledge about the possible security vulnerabilities also contribute. With these areas as the focus, statistical modeling for software source code tries to derive the knowledge hidden in various software artifacts including code and text and help to build a robust and intelligent system. Integrated Development Environment (IDE) is one area of focus on the subject. Since this provides a platform for entire software development processes, it plays a key role. In this chapter, study of intelligent IDEs is extended to the IoT environment context. To begin with, we focus our study on understanding the IDEs and their capability. Exploration will focus on open-source IDE for mobile, and we look for a better understanding of the landscape so that we can extend the learning to the IoT world. In this work building, low-cost IDE for mobile is focused open. Also, interestingly, exploration focuses on open-source components and the possibility of putting them together. Then, in the next part, we explore the IDE for the Internet of things (IoT), with a focus on the open-source ecosystem. In this review, we extend the exploration of IoT into device management, data management, communication, intelligent data processing, security and privacy, and application deployment areas. This gives a greater insight into the IoT world to extend the need for intelligent IDEs to the IoT world. Then, to get the context of IDE in the machine learning context, we explore the topic of building an optimal IDE for feature engineering which is one of the key phases in the machine learning life cycle. Since machine learning projects are a highly data-oriented ecosystem, learning of IDE and its insights in this area will provide rich insights into the main theme of discussion which is an intelligent IDE for the IoT ecosystem.

2 Overview of the Integrated Development Environment

Integrated Development Environment (IDE) eases the development of the software, facilitating the usage of multiple tools from a platform of development. IDE for a device would become expensive when we look at an embedded system, this is particularly due to the number of resources involved. In this review of the work, we look at mobile device integrated development environment development with Internet tablet Nokia 770 taken as reference. Work has targeted low-cost full-fledged IDE. Interestingly, open-source components are explored and integrating of the components into practical and simple IDE is targeted at. Most parts of the applications are build using some form of IDEs. The application package for installation and cross compiler makes IDEs usage in embedded systems development more important. Since the life span of the embedded device is short and they are specialized device in most of the cases, the cost associated with the investment for the specific tools will become costly. This makes the reusability of the components a more preferred option. The existence of extensive code and adoption of the code for various purposes make open-source software most suited for the situation. The main idea of the experiment here was to have a software development tool with an integration of several open-source projects where the provision of using the components of the system be made seamless. Eclipse [1] was used as a base platform to build systems on top of it. Eclipse provides a common vendor development platform that helps to build software with frameworks for application. Eclipse having the plugin architecture was the choice and this provides the possibility of extending the same to different needs. This feature of the development environment that is language agnostic, will give a great base for an integrated development environment. C/C++ development tool kit (CDT) provides the plug-ins for Eclipse. These CDT tools provide simplified tooling that can be used from the command line. External utility communication and interpretation of the responses are the features of the CDT that comes in handy. CDT plugin source code was explored to see its potential to operate in a sandbox environment. Laika open-source software is involved. Communities related to Laika are CDT, Eclipse, Gazpacho, Maemo, Pydev, Scratchbox (Fig. 1).

Another development environment Scratchbox [2] which is a cross-compiling tool is used for embedded devices development for Linux desktop. Scratchbox provides

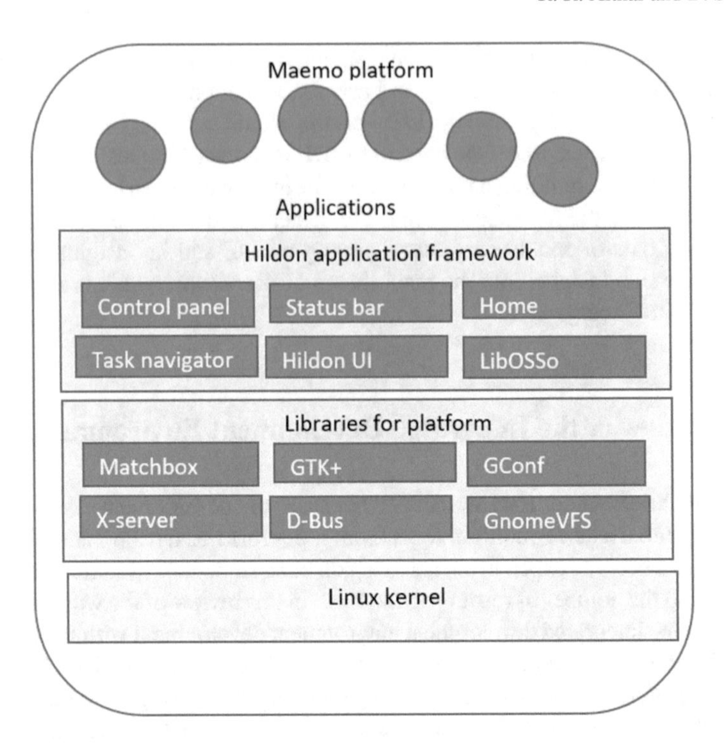

Fig. 1 Maemo platform architecture

as its 'targets' different configuration of CPU architectures. For the different target of the platforms, different configurations can be applied, various toolkits can be applied to these targets as well.

Using Scratchbox in IDE and CDT features expansion was the lookout of this study. Parsing, debugging, launcher, and syntax highlighting are the key features of CDT. A lot of the other features on top of CDT features was tools for automatic package creation and utilization of auto make tools. Application templates were also targeted in the experiment (Fig. 2).

Key learning from this part of work was, without comments expansion of the large-sized plugin was a tough task and due to IDE integration in Scratchbox CDT operations inside it was good but had led to confusion at the beginning of the experiment. Stabilization and improvements were the focus of the next phase with the stabilization of naming convention and package conventions. Learnings from here were the co-ordination that was needed in case of teams that were operating in different geographical locations, without which the redundant work led to additional time involved. The later focus was at integrating the Python to Laika, as it was decided to be used as the main scripting language, and this was possibly explored to showcase the potential flexibility that the platform had to offer. Various plugins of the Python were targeted to be integrated. With the Python plugins, there was a rapid change in the versions, but since the open developers' community was engaged,

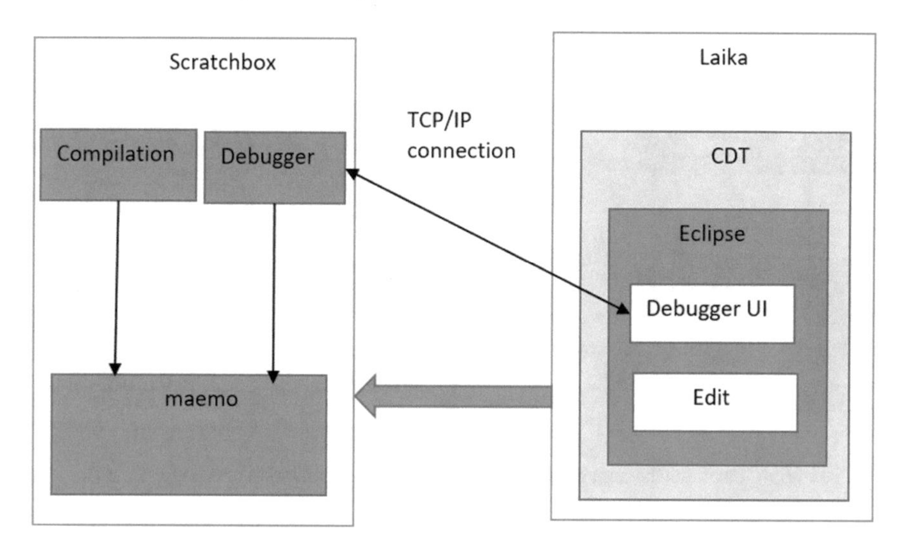

Fig. 2 Components and their relations with each other is illustrated in the figure

it made it easy to employ those changes. Apart from the version changes involved in the plugin, the technical challenges associated with the development was also to be handled. Compared to CDT tools, Pydev [3] debugger, one of the plugins used for Python had frequent communication with Eclipse. This is complicated due to Scratchbox as a sandbox usage will be tough as Pydev houses the Python debugger inside it unlike the case of CDT where the debugger was part of the IDE. Setting up Python interpreter was another major issue; ARM interpreter was challenging, as it would not be accessed from Eclipse as binary were ARM executables. Also issue with the failure of Pydev source code in Sourceforge, when tried to download the same. This led to have Pydev class files with parameters and function names only developing blindly without source code (Fig. 3).

Key take away at this point was, code comments were very low and for code delivery, Sourceforge was not enough, to be stable. The practicality of having collaboration with the key developers from the community was realized to be key. It's a great resource optimizer to bank on existing open-source resources. Lessons that have come out would be great input while working in Integrated Development Environments. While we are looking for compatible components, the lack of information gathered for components of the open-source will be challenging. The inadequacy of the comments involved in the open-source plugin will be challenging. The short period introduction of changes in the third-party components makes it challenging to handle. Integration complexity spikes up as embedded systems component count goes up. Consistency in the way the open-source components are used within the IDE is not seen, and this leads to difficulties. Even in case if there is clarity of using components consistently, streamlining it across the community will be challenging. Though the challenges are quite significant, some effort that can be considered would

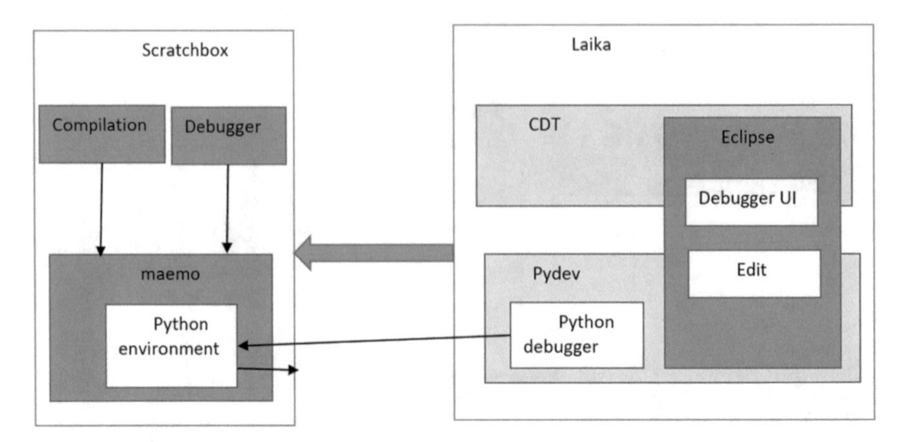

Fig. 3 Figure illustrates the next stage of the experiment at stabilization and improvements

be. Better to start with a short-term plan rather than targeting all the problems. Subsystems integration to be strategized well. The selection of the partnering community is the key, as the scarcity of the developer's community involved may cause serious hamper. Engaging with too many communities becomes troublesome in the case when there is a large number of components introduced and not all communities succeed to adopt the same. Also getting accustomed to such a large community will ask for time investment. Putting the experiments into practical use will help understand the situation better and learn faster. Open culture to engage all the developers involved and platform for users to provide the input will contribute [4].

3 Open Source Ecosystem for IoT in an Integrated Platform

The Internet of things has become a key part of the life of people. Complex management, heterogeneity, and other difficulties are bugging the development of Internet of things (IoT). IoT-integrated development platform architecture will be looked at. Advancements in the software and hardware technologies have enhanced the capabilities of the IoT to get into the life of people in a more granular way. Monitoring data from sensors go into storage on the cloud computing platform on the network and leads to intelligent processing, the respective environment is modified to meet the need of the people. To combine the real need of the IoT applications, extensive research is on. Web of things testbed (WOTT) [5] proposal intends at simplification of design and development environment with an open, secured, and flexible environment that is stable. The micro-service architecture proposed in the platform [5–9] which intends at having an open-source system for integrated application development which provides end-to-end development and deployment environment integrated for the applications. Practical integrated application development environments for the IoT is the target for this work. Existing experimental platforms do not account for standardization with real scenarios, which needs focus. Intelligent data processing for multiple sources, service development personalization of the IoT, various applications, and heterogenous device support are the intended areas of the work.

4 Integrated IoT Application Development Platform Architecture

Power consumption, protocols for transmission, and operating systems of sensing devices functions are supported by various devices that operate on heterogeneous activities. Application requirements of the current mainstream requirements are targeted at. Heterogenous devices management with the gateways is done. Forwarding, storage, unified description, basic processing of data and other activities are implemented by the gateway. Device location identification also is possible. To establish integration among various applications, protocols for application layers like HTTP and Advanced Message Queuing Protocol (AMQP) to achieve data transmission in application layer protocols. To meet real-time requirements of the applications and to ensure loose coupling in the system, subscribe mechanism is also inactive (Fig. 4).

The sensing device is based on the application development of IoT, which provides resource management in a distributed way. Specific functions and features are supposed to be used in the development of the devices. So common public service or discovery mechanism for the resources provision becomes important to

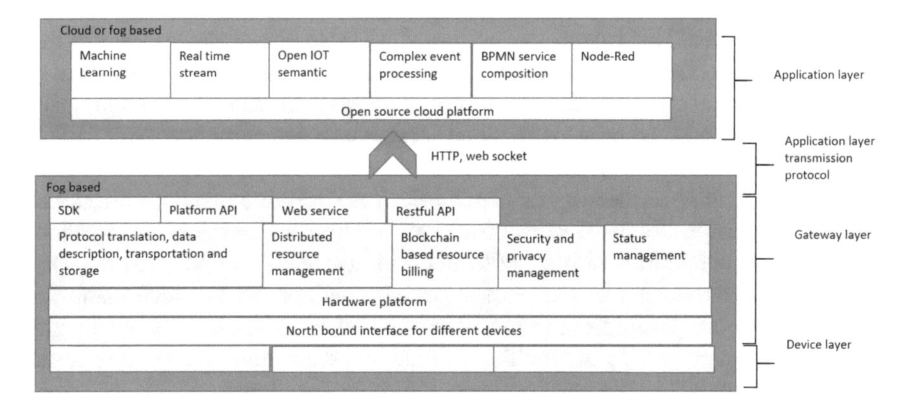

Fig. 4 Open-source technology-based application development platform for IoT

make resource discovery smoother. For virtual currency trading and sharing of the resources, peer-to-peer (P2P) computing technology which is distributed is extensively used. Retrieving of IoT resources and distributed storage is the platform offering by making use of P2P. Flexibility, scalability, and high availability are the specializations of IoT application with P2P. RESTful API and Web services are offered in a platform to facilitate API interface with the security mechanism. Distributed Ledger Technology (DLTS) and Transport Layer Security (TLS) mechanism for embedded system ensuring network layer security providing authentication for identity and encryption of data. For secure transmission of the information, Jason Web Token (JWT) is utilized. Data integration is managed with the signing of the transferred information. With options for development libraries, continuous development environment integration, API provides a build system with cross-platform, this is provision for service delivery-based platform with the environment for development. The cloud platform provides the provision to development and deployment of the application of IoT. Information on a variety of sensors is stored by the platform in the cloud. These data are further leveraged for intelligent knowledge processing in areas of machine learning and data mining. The platform consists of Apache Kafka with cloud platform Eclipse Kapua [10] for the IoT data processing platform. The efficient and safe use of sensor resources was ensured with resource bills from blockchain technology. Smart contract technology with blockchain helped manage resources for IoT.

5 IoT Application Development Platform Features

The interoperability module was which is part of a middleware reference model that was proposed by Da Cruz et al. [11]; it provided the resources for integrating, data processing and storage module, relevant information providing context module, for event generation and management event module is part of it. About this setup, the architecture of the system that is proposed in this work discussed here has the following features. The system can be hosted close to sensor deployment in the fog computing model. This provisions local data computing and avoids transfer of the data to computing point which are remotely hosted. Transmission of the transformed data would then be done to cloud servers to make computing an efficient collaboration between the source of data and the cloud. Multiple devices are made compatible shielding the heterogeneity of the devices with northbound interfaces and make them scalable. For development purposes, southbound interfaces are provided. This helps by supporting various API interfaces like RESTful APIs. In the case of a host of subsystems like a case of smart city, multiple open-source platforms and self-developed platforms are adopted to use Web of Things technology. Intelligent application processing capabilities are built into the system with context-based processing capabilities with business logic and application layer.

6 Evaluation

Industrial IoT scenarios were used to validate the operations of the platform discussed here. For safety in production and efficiency, for the industrial setting data acquisition process, management of the devices and monitoring of the system can be implemented with IoT technology. Production environment with sensors picking up the data for various process monitoring is an industrial setup in most cases. A variety of devices and transmission needs, complex data processing, and complex need of real-time data processing are the need for a smart industrial setup. In this setup also the abnormalities are figured out by the cloud platforms that receive the sensor data from the locations. Factory setup also are at remote places from the management folks. This makes it important for remote data collection and processing and wireless mode in many cases. Enterprise applications can get these setup easily (Fig. 5).

Application-oriented protocols assist in easy integration across applications. Remote data transmission is facilitated by HTTP protocol with Constrained Application Protocol (CoAP). The facility provides the ability to add on more application layers. Evaluation is accounted for system function implementation for application development in an integrated fashion. The network environment of the campus was considered for evaluation, more complex setup is planned to be used in the future. The interoperability of gateways from multiple systems is said to be the focus of future evaluation, whereas the current one is focused on a common gateway. IoT gateway management was done on the Eclipse Kapua cloud computing platform that provided system configurations, data publishing, and subscriptions. For system performance analysis, of IoT platform covering connectivity, performance, functional, security, and exploratory analysis, Esquiagola et al. [12] proposed testing methodology. Various mainstream application layer format was used for performance comparison. IoT application developed by Node-Red runs on the Node-Red browser. Gateway and IoT application communication is figured out with Round Time Live

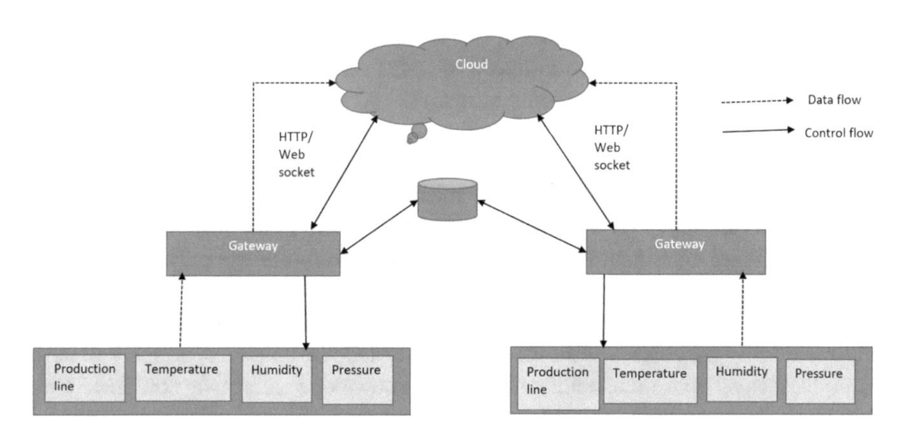

Fig. 5 IoT simulation in an industrial setup, with management control for acquisition and monitoring

(RTT). Request time (T_{req}), request processing time (T_{p}), time of request for response (T_{res}), time is taken to display the data received (T_{dis}) are the various data involved.

The below equations depicts the performance tests conducted.

$$\mathrm{RTT} = T_{req} + T_{p} + T_{res} \tag{1}$$

$$\mathrm{RTT} = T_{req} + T_{p} + T_{res} + T_{dis} \tag{2}$$

Temperature, humidity, and other data in case some small setups are small-sized, but in case of larger data processing setup, the performance of the network may get impacted due to more frequent data being transferred, and in those cases, there should be multiple attempts for data transmission. This evaluation is conducted with testing carried with various payloads. PUT operations were changing the state of the resources whereas GET operations obtaining the resource data.

6.1 Summarization

Integrated application development platform architecture with cloud-fog combination based on the open-source was discussed. This setup was evaluated in the industrial setting. For resource billing management, smart contracts with blockchain were explored and P2P technology for resource management in a distributed way was discussed. Architecture validity was provided through initial function tests. Open-source ecosystem-based IoT gateway was tested with gateway test and proved to be reliable and stable, and this was proved for the data size and concurrency which was suitable for IoT in most of the environment where the sensors were applied. More functional implementation is to be explored in further research. Preliminary implementation with performance analysis was done with Eclipse Kapua cloud computing and Kura middleware for application development platform with a simplified topology. Future exploration needs to focus on the utilization of middleware open-source collaboration environment addressing complex scenarios. For LoRA network environment with HTTP network performance for communication of long-distance, further analysis and comparison are needed. Also, further study should focus on automatic access, registration, release, an update of the physical entities as part of device lifecycle management. Data processing and customized services platform will get intelligent technologies integrated as future work. To use data that is collected by IoT in a more intelligent way, machine learning adoption will be one such example of data processing intelligence that can be built-in. For application interoperability purposes, useful information needs to be collected, which will have a rule-based system for reasoning purpose and knowledge base being combined with this. For users to enjoy personal IoT service provision efficiently, Software Defined Networking will be adopted.

6.2 Faster Feature Engineering for IDE

So far, we explored the background of the Integrated Development Environment (IDE) and its application in the context of the Internet of things. Now its time to explore how the concept of IDEs has been looked at in other contexts, which provide more insight into extending its application in the world of IoT [13]. For many software system developments, utilization of a large amount of data and running machine learning algorithms on them is a key aspect these days. All the experience of big companies like Google, Netflix boils down to the fact of engineering the features. Machine learning system performances are influenced by good quality features that are built with great difficulty and a lot of time expended. So, to facilitate this area developer tool that can uphold the feature engineering will be handy. For the feature engineers, this work reviewed here has an Integrated Development Environment. This particularly comes in handy for evaluation of the effectiveness of the feature code that is important. To identify the relevance of the data to feature code for the user, raw data is processed with the execution planner at runtime. IDE particularly helps for arbitrary code for feature and looks at its influence in learning the pattern and keeps tab of pace at which approach can be applied when compared to baseline performance. Extracting data from the very large dataset is a key underlying process that happens in the case of IBM Watson's question answering system or Netflix's recommender system or the search engine of Google. Even with these big giants, these are the system build over years of dedicated effort. Ultimately, this difficulty points back to feature engineering.

Features are the extracted pattern from the raw data which become the input for the machine learning system in the supervised learning setup. To capture the relevance of the user query to the searched content, a Web search engine-based regressor which is trained is in play. In this context, relevant features will be like, if the user query comes as part of italicized content or part of document title and so on. For generating human-built features, the feature code defined by the user is applied to the raw data. Good feature code will help to generate valuable features for machine learning tasks and will be relevant ones from a software engineering perspective also.

Good features play a major role in the performance of the systems that are trained [14]. Also, interesting facts are that the good features are built over period and less so with a breakthrough. It is not easy for engineers to derive features as the raw data are extensively large-sized and the feature specifications are not very clear. Adding to the complexity is the information of whether the feature helps to improve the accuracy of the system. There would be scenarios where the programmer may implement the functions with great difficulties to realize that the feature may not work well. So usually engineers conduct multiple rounds of iterations which are small and informative to incorporate the feedback less expensively. Each of these iterations also will be time-consuming, as the function associated with the feature must be applied to the large dataset for training and similar effort needed for testing purposes. This work targets to expedite this process of iteration. This target improves the productivity of the engineers with a faster compilation time though it may not help to improvise the quality of the code. Thus, the contribution is on part of productivity to make the system less expensive and accurate.

6.2.1 Feature Engineering Today

Most of the software bundles do not provide support for feature engineering. For feature code development, the evaluation sequence used are, MapReduce [15] bulk data processing systems help in case of applying feature functions to raw data that are large enough. Though this task consumes time, the results will be desired feature vectors being generated. These features will be used to train the machine learning models with software toolkit such as Weka [16]. Based on the technique of machine learning chosen, run time cost will be decided. Accuracy of the results is evaluated on the dataset that is held out for this purpose, an example can be text classifier that is trying to classify article into the sport, politics so on. Based on the need for accuracy, the code for the feature can be refined and the steps involved can be regenerated. When the required accuracy is reached, then the job is completed. Optimization of the runtime cost is a prominent area of research. This work reviewed here focuses on the initial step of processing large data. Current systems lack due to the data getting processed in a random fashion and most case full set of data getting processed. In case if this part can be made efficient by choosing to work on the beneficial part of data, then as we hit the right results experiments can be cut short.

6.2.2 Challenges

The central part of the experiment will be about more meaningful processing of the data, so optimal selection of input data will be needed. Rather than processor operating on all data, the choice of data that can make the best use of users' feature code can be looked at. To take an example is if the feature code works well with the news related information then that information needs to be picked. Likewise, e-commerce prices accustomed to data will look at pages from the e-commerce page. How useful the features are can be variable as the feature code keeps changing with every invocation of the system. In this context, it is tough to pre-process input data and make input usable, so the system needs to do on the fly identifications of relevant input. This relates to the active learning scenario where the learning algorithm allocates the data needed for learning purposes in a supervised setting [17]. For improving the accuracy of the artifacts, suitable training features need to be selected by the system to maximize improvement. It is a choice in this setting to avoid the cost associated with the time when compared to active learning. This feature of active learning makes it irrelevant. Domain independent indexing with a planner at runtime is the technical solution involved here and is called Zombie. Authors were on writing different papers to explain technical details of this system Zombie, here working IDE built on Zombie is discussed. The system is devised to take as input user feature code and apply these functions on the large dataset related to Web pages. A Zombie system or standard baseline system for data can be used by the user for data processing purposes. Amazon EC2 services with a configurable number of machines are used to run the tasks. We will focus on the evaluation of the architecture of the system that works on feature code, also explore the user interface.

6.2.3 System Framework

Input and output of feature development system are to be explored. Overall learning task configuration with necessary parameters is to be done by the users in feature IDE, in addition to the novel feature code addition. In the case of the raw dataset such as Web, crawl feature vectors are derived by IDE from the dataset using feature code. Distant supervision technique [18, 28, 29] is used to obtain human-generated features; for generated feature vectors this method provides supervised labeling. Learned artifacts produced by consuming the featured vectors after labeling with the training procedure adopted by the machine learning algorithm naïve Bayes classifier are one such example. The machine learning library provides this option. Hold out a dataset is used to assess the accuracy of the artifact after training; this uses a method for this scoring purpose for the quality measure. Feature code engineered will undergo frequent changes. Parameter discussed above will remain constant throughout the process of engineering tasks. These parameters also decide the environment for user feature code success. Four-way classification tasks are configured by these parameters with prior configurations done with Wikipedia data, semantic tags of the pages are used for obtaining the labels and use multi-class naiveBayes classification [30, 31]. Zombie system behavior is fully utilized for this scenario that is designed with a simple task for users to get a full evaluation. IDEs main interface is initiated when the user is ready with the feature code. Every element in the raw data set then gets its application of feature function code from the development system of the features. This step of feature function code usually is time-consuming since the dataset is very large. The Zombie system is applied by the IDE in this case, or MapReduce would be used by a standard feature development system. By just running with the available feature vectors and turning off the function in the application, engineers reduced a considerable amount of time. In that case, the baseline system will have many of the vectors of high value lost from the output [32, 33]. Since, in the case of Zombie, the selection of useful features happens, early terminations of the function will not cause much of the loss. When compared to the training vectors produced blindly by the MapReduce, Zombie-based IDE can do better production of trained vectors with the available amount of processing time. This makes it transparent that if the run is completed, the output produced by the baseline system and Zombie will match (Fig. 6).

The above visualization is for a single run and a single set of feature code written by the user on IDE. Raw data item processed of high value is shown in the Y-axis and time for feature function run is shown in X-axis. In the dataset used here, the high-value datasets were up to a range of 0.5% unlike in the case of real-life, these high-valued ones will be much lesser. This is since the target data is coming from

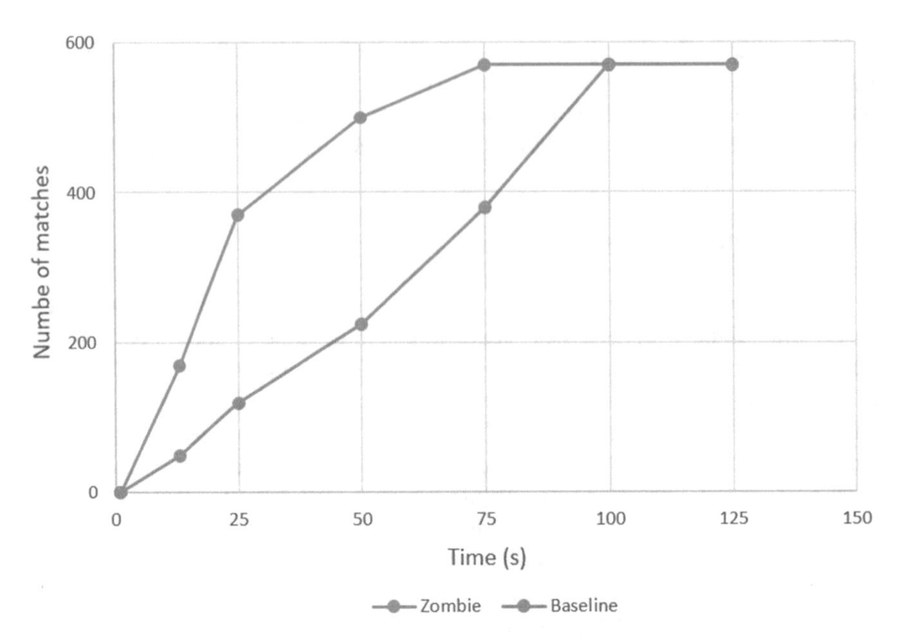

Fig. 6 Zombie and baseline performance comparison for matches over time

the Web pages of the faculties in the computer science department. High-value items of 0.001% can be obtained when Web page crawled is about five billion pages for an estimated fifty thousand faculties across the USA. Even if the extension is made across academic institutes, this share may go up to 1%. Unnecessary simplification is involved in this classification of the inputs into high and low values. The fine-grained difference in their use is also a characteristic feature of the data used here. Based on whether similar pages have been seen before the utility of the pages will be decided. As the utility of the faculty from an unseen institute will be of higher utility when compared to the institute from which more inputs have already been utilized [34]. Visualization seen above is time-based IDE keeps updating the data points as the run progresses. This time-based monitoring will help engineers to take a call when to terminate the function after assessing that the good quality function vectors are generated. Visualization indicates that the Zombie system produces three to four times more high-value features in lesser time compared to the baseline system, and this was possible with the time of say about 25 s. At about 75 s are so both systems find an identical number of items. It is notable that in general setting, these time frames may run up to several minutes or even hours.

6.2.4 Approach

The Zombie system for execution provides the core functionalities for the system under discussion here. A key feature is to identify the feature vector that provides high value without spending resources on computing the feature vectors from learned artifacts. Multi-armed bandit problem [19] has a similar context as seen here, where the agent must keep a tab of the outcomes paid by every round of the execution to assess the best paying entity. In this context of multi-armed scenarios for the agents, there are optimal policies that are configured, in the Zombie system [23–25]. Reward functions and set of bandit's arms to pull from are the key components of bandit configuration. The pull of the bandit's arms will follow with the reward associated with the arm. Initial clustering of the raw data is obtained from arms combinations that are not related to the feature code. A planner from Zombie will randomly decide to pull the arm from the cluster and use that to process with the feature code devised by the programmer. The impact of the resulting feature vectors, from the reward of the pull, on the quality of the learned artifacts is in play. To select the next arm or the cluster of the inputs, Upper Confidence Bound Algorithm (UCB1) [20–22] is used.

6.2.5 System Architecture

Web application and engineer's browser are in communication while the code for features is written. Amazon EC2 service is utilized by IDE when an evaluation of the feature code is started by the engineer. The usable feature vector is obtained by applying the user feature code on the raw data utilizing the worker machines that are generated for this purpose. Unlike in the case of baseline, workers just scan through complete data [26, 27]. Feature independent indexer utilized helps Zombie workers to screen through for the best feature vectors on priority. The master node is reported with the best features at regular interval of time of seconds, by multiple workers that are focusing on different areas of the raw data. The user browser gets the consolidated statistics from the master node which has put together all the information received from the nodes. In this experiment, baseline workers are included for comparison purposes only, but in real-life examples, the resources would be optimized with the Zombie workers only. Though the experiment intends at the intelligent selection of the high-value feature vectors, lookout is also toward building good quality learned artifacts. This being the context run time statistics become a key feature (Fig. 7).

6.2.6 User Interface

Two primary pages are available on the Zombie IDE Web application. Users can choose from the list of pre-written functions or write code from beginning on the IDE code editor. Code will be related to feature code and configuration of related tasks can be done. Ideal usage of the system would see that the engineer will end most of the time using this code editor. Authors of this work were planning to add more features to the editor of IDE based on the work from Anderson et al. [14].

There is also a runtime control panel that the engineer can use for evaluation purposes of feature function application. Also, on this console, users get control over the job to choose several machines needed and control for starting and stopping

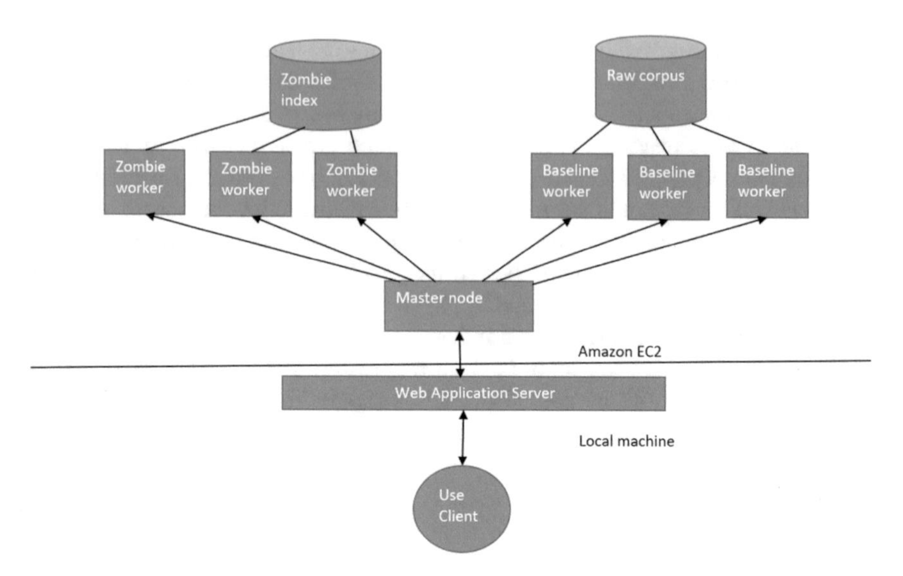

Fig. 7 Run time architecture of IDE. Interface to the user is provided by front end Web server which creates on the machine the Amazon EC2 service for the needed number of services

the system. Overall, the interface has the view of Zombie's display of the ability to identify the inputs of high value. This part is where the live performance happens which we monitor in the visualization that was discussed earlier. This performance data of the visualization keeps updating every few seconds. Based on monitoring done on this data, users will be able to take a call on when to terminate the function. A provision in the system provides the ability to modify the feature code in the editor regularly and then in run time, control panel evaluates the code-switching into it.

6.2.7 Demonstration

Real text classification tasks can be tested on zombies by the users that intend to test the demonstration. For users to get the view of the performance in quick time unrealistically, small data of the corpus is provided as input data and interaction is demo styled. Performance checks by users will be between zombies and baseline systems. In the case of testers want to write their feature code then the following details illustrate how IDE provided here can make this process more meaningful. Say, users want to build a classifier of the article and want to classify the news Web pages into politics, sports, and entertainment. With Wikipedia pages as the raw input, for users to make changes in feature code lot of time must be expended as this is an iterative process, also the evaluation associated with the change of the feature code will add on to the time. In these scenarios, unclassified category will be a larger part and Zombie should fix its priority on the categories that have classes defined. Two parts evaluation is conducted; the quick first part just looks for the existence of keywords related to the topic that is looked out for [35, 36]. The second part is more complex in the sense it must conduct natural language parsing of the content. These tasks are applied sequentially, and the baseline and Zombie are run in parallel. Compared to baseline, Zombie will bring up value-adding features at a quicker rate though there is some amount of system overhead. As part of evaluation, users will see that the benefit in case of the second function, that does data parsing, being more advantageous compared to the function that looks for key terms. Users also will have the provision to modify the feature code and see the difference in the performance of baseline and Zombie. As the testers define their function, these become input for the research that the team is conducting.

6.2.8 Summarization

The entire discussion was focused on bringing out the advantages of building a data-centric IDE; this should pave the way to further refinement of these methods and utilize the knowledge in a variety of situations.

6.3 Intelligent IDE for IoT with Statistical Source Code Modeling

6.3.1 Probabilistic Programming for Statistical Source Code Modeling

Probabilistic programs are including a part where random values are drawn from distributions for inferencing purposes. These distributions are the outcome of a study of the data patterns. A second key part of the probabilistic programs is the conditioning of the variables involved in the program with the observation of the data. Examples of the probabilistic program applications are in areas like reliability analysis, coding theory, and computer vision. Inferencing involved in the process is probabilistic and has its base in probability distribution derived from the data, which is processed by the probabilistic program. Outcomes under review will depend on the scenarios involved; in general, expected value to be found for a function about the distribution of the complete data or sample drawn from the data. Work here focus on reviewing the connection between probabilistic programming and software engineering programming language, extending to language design and program analysis covering the dynamic and static part. As part of probabilistic programming, the Bayesian approach that includes the theory of conditioning will hold the key. Automation of the Bayesian inference in probabilistic programming. In the context, it is about generating semantics and syntax for the language, for problems that need conditional inferencing [37]. Probabilistic programming is about statistics done with computer science tools. Computer science has been the area focusing on the efficient evaluation of the program consuming the parameter and arguments to generate the expected output. Probabilistic programming system inferencing focus on designing the algorithm that can be used for any program that users intend to use and in any language. Joint and conditional probabilities are involved in the case of probabilistic programming; in the case of conditional probability, there is a syntactic indication of where the condition is applied. Everywhere in most languages, there are pseudo-code generators, but probabilistic programming stands out due to their constructs that are syntactically based which work on implementing the condition by conditioning and evaluation.

For purpose of graphical modeling and inferencing, variety of libraries are available, model representation in case of these would be programmatically construction of the data structures, based on these models graphical inferencing is run. The difference between these graphical models and the one of probabilistic programming is, the graphical model uses programs to construct data structure-based models whereas in other case, models are devised based on direct evaluation of program expression [38]. Also, in probabilistic programming focus is on either building the model based on data structure with the interpretation of the probabilistic program or other approach is to use the Markov model where the state keeps evolving in the environment. Probabilistic programming has its key application in making the outcomes more readable when compared to mathematical notation, for the model. Procedural graphics is one of the key areas, that involves stochastic simulation, with application like computer graphics forest simulation. A stochastic simulator generates a group of trees by calling the program multiple times.

6.3.2 Key Areas of IoT

After the exploration of the IDE in the context of mobile devices and the utilization of open-source components to make it cost-optimal. Then, looking at the IDE in the context of the IoT, further exploring the optimal IDE setup for the data-rich machine learning context. Now, we look at exploring the possibilities of using knowledge of software source-code statistical modeling in the area of IDE for IoT. IoT opens the challenge of handling data from across the multiple sources of sensors, handling this data centrally and facilitating the intelligent processing of the data will be an area to look at. A secure and open environment for IoT design and development is an area that has been explored and needs some attention. Micro-architecture proposals for the integration of the IDEs would need the knowledge for statistical modeling of the source code. Secure data transmission is another prominent focus area that can be managed. Since Cloud platform is involved, it will be important to monitor the resources and ensure optimal utilization and manage cost, and this scenario is also appropriate to handle. The trade-off between computing at the device end and cloud will be a good area to facilitate. Real-time monitoring and reporting need can be managed effectively. Abnormalities detection and management with the prediction of the life of the devices are involved. Things in the physical world and the virtual world need to get connected in a communication network that enables the opportunity for efficient integration of the system. Heterogenous components involved in the IoT space covering software and hardware also are the key area to focus. The expectation of the common software functionalities to be able to be deployed in multiple systems with only small modifications calls for efficient implementation of the automation as well as handle complex things. Need for streamlining the communication between various stakeholders involved in the system. Towards making productivity a key focus, auto-generation of the code, auto-completion, and other related areas are open. The abstraction of the models into different lower-level versions to establish the impact of changes at any point in the software development life cycle. Adaption to the changes will be easier from the business and technology point of view when there is a model representation of the development systems. On the IoT systems, management of heterogeneous technology on a common platform is a key area. The theme of reusable components in the IoT, in a system built with the inspiration of the service-oriented architecture, forms another focus area. The middleware of the IoT systems, that is formed out of the vast number of components, variety involved in the data and their extensive size, the different patterns needed for the data involved are the prospective areas for exploring statistical source code modeling.

References

1. Eclipse, "Eclipse". Available at https://www.eclipse.org. Visited 6/5/06.
2. Scratchbox. Scratchbox—Cross-Compilation Toolkit Project. Available at https://www.scratchbox.org. Visited 06/05/06.
3. Pydev. Python Development Environment. Available at https://Pydev.sourceforge.net/. Visited 7/5/06.
4. Järvensivu, J., Kosola, M., Kuusipalo, M., Reijula, P., & Mikkonen, K., Developing an open source integrated development environment for a mobile device. Institute of Software Systems, Tampere University of Technology, P.O. Box 553, FI-33101, Tampere.
5. Belli, L., Cirani, S., Davoli, L., Gorrieri, A., Mancin, M., & Picone, M. (2015). Design and deployment of an IoT application oriented testbed. *IEEE Computer, 48*, 32–40.
6. Gordon, A. D., Henzinger, T. A., Nori, A. V., & Rajamani, S. K., Probabilistic programming.
7. Fit IOT-LAB. Available online: https://www.iot-lab.info/. Accessed on June 18, 2018.
8. Sotres, P., Santana, J. R., Sanchez, L., Lanza, J., & Munoz, L. (2017). Practical lessons from the deployment and management of a smart city internet-of-things infrastructure: The SmartSantander testbed case. *IEEE Access, 5*, 14309–14322.
9. Sánchez, L., Gutiérrez, V., Galach, J. A., Sotres, P., Santana, J. R., Casanueva, J., et al. (2013). SmartSantander: Experimentation and service provision in the smart city. In *Proceedings of the 16th International Symposium on Wireless Personal Multimedia Communications (WPMC)* (pp. 1–6), Atlantic City, NJ, USA, June 24–27, 2013.
10. Eclipse Kapua. Available online: https://www.eclipse.org/Kapua/. Accessed on June 25, 2018.
11. Da Cruz, M. A. A., Rodrigues, J. J. P. C., Al-Muhtadi, J., Korotaev, V., & Albuquerque, V. H. C. (2018). A reference model for internet of things middleware. *IEEE Internet Things Journal, 99*, 871–883.
12. Esquiagola, J., Costa, L., Calcina, P., Fedrecheski, G., & Zuffo, M. (2017). Performance testing of an internet of things platform. In *Proceedings of the 2nd International Conference on Internet of Things, Big Data and Security* (pp. 309–314), Porto, Portugal, April 24–26, 2017.
13. Anderson, M. R., Cafarella, M., Jiang, Y., Wang, G., & Zhang, B., An integrated development environment for faster feature engineering.

14. Anderson, M., Antenucci, D., Bittorf, V., Burgess, M., Cafarella, M., Kumar, A., et al. (2013). Brainwash: A data system for feature engineering. In *CIDR*.
15. Dean, J., & Ghemawat, S. (2004). MapReduce: Simplified data processing on large clusters. In *OSDI*.
16. Biswas, J., Kureethara, J. V., Samanta, D., & Sandhya, M. (2020). Efficient algorithm for people management in an elevator. *TEST Engineering & Management, 83*. Publication issue: March–April 2020, ISSN: 0193-4120.
17. Samanta, D., Galety, M. G., Shivamurthaiah, M., & Kariyappala, S. (2020). A hybridization approach based semantic approach to the software engineering. *TEST Engineering & Management, 83*. Publication issue: March–April 2020, ISSN: 0193-4120.
18. Manral, S., Samanta, D., & Podder, S. K. (2020). Effective classroom activities on accounting using double entry system: The productive consequences. *TEST Engineering & Management, 83*. Publication Issue: March–April 2020, ISSN: 0193-4120.
19. A Khamparia,P K Singh,P Rani,Samanta, D,A Khanna, B Bhushan. (2020). An internet of health things driven deep learning framework for detection and classification of skin cancer using transfer learning. *Transactions on Emerging Telecommunications Technologies*. ISSN:2161-3915 2020.
20. Gurunath, R., & Samanta, D. (2020). Studies on encrypted secret data storage techniques analogous to steganography. *International Journal of Advanced Science and Technology, 29*(2), 3705–3711.
21. Li, Y. Q. (2018). An integrated platform for the internet of things based on an open source ecosystem. *Future Internet, 10*(11), 105. Received: September 21, 2018. Accepted: October 30, 2018. Published: October 31, 2018.
22. Roy, S., Kanti, M. M., Samanta, D., & Venkatanagaraju. (2020). Awareness with informatics on hypertension and effects on hemoglobin. *International Journal of Advanced Science and Technology, 29*(4), 423–433.
23. Samanta, D.Sivaram, M., Rashed, A., Boopathi, C..S., Amiri, IS., & Yupapin, P (2020). Distributed feedback laser (DFB) for signal power amplitude level. *Journal of Optical Communication,*doi: https://doi.org/10.1515/joc-2019-0252.
24. Gomathy, V., Padhy, N., Samanta, D., Sivaram, M., Jain, V., & Amiri, I. S. (2020). Malicious node detection using heterogeneous cluster based secure routing protocol (HCBS) in wireless adhoc network. *Journal of Ambient Intelligence and Humanized Computing*.
25. Thomas, B., Shwetha, P., Dey, P., Biswas, J., & Samanta, D. (2020). An efficient and holistic approach to reduce output and dependent parameters for multi-output learning. *International Journal of Advanced Science and Technology, 29*(4), 25–33.
26. Chatterjee, K., Samanta, D., & Biswas, J. (2020). Enhancement of education with wearable computing device. *CSI Communications, 43*(10). ISSN 0970-647X.
27. Anwar, Z., Banerjee, S., Eapen, N. G., & Samanta, D., A clinical study of hepatitis B. *Journal of Critical Reviews JCR, 6*(5), 81–84. https://doi.org/10.22159/jcr.06.05.13. E-ISSN: 2394-5125.
28. Kureethara, V., Biswas, J., Samanta, D., & Eapen, N. G., Balanced constrained partitioning of distinct objects. *International Journal of Innovative Technology and Exploring Engineering*. ISSN: 2278-3075 (Online).
29. Sivakumar, P., Nagaraju, R., Samanta, D., Sivaram, M., Hindialraj, M. N., & Amiri, S., A novel free space communication system using nonlinear InGaAsP micro system resonators for enabling power-control toward smart cities. *Wireless Networks—The Journal of Mobile Communication, Computation and Information*. ISSN: 1022-0038 (Print), 1572-8196 (Online).
30. Samanta, D., & Podder, S. K. (2019). Level of green computing based management practices for digital revolution and new India. *International Journal of Recent Technology and Engineering, 8*(2). ISSN: 2277-3878.
31. Mahua, B., Podder, S. K., Shalini, R., & Samanta, D. (2019). Factors that influence sustainable education with respect to innovation and statistical science. *International Journal of Recent Technology and Engineering, 7*(5S2). ISSN: 2277-3878.
32. Praveen, B., Umarani, N., Anand, T., & Samanta, D. (2019). Cardinal digital image data fortification expending steganography. *International Journal of Recent Technology and Engineering, 7*(5S2). ISSN: 2277-3878.

33. Manu, M. K., Roy, S., & Samanta, D. (2018). Effects of liver cancer drugs on cellular energy metabolism in hepatocellular carcinoma cells. *International Journal of Pharmaceutical Research, 10*(3). ISSN: 0975-2366.
34. Paul, M., Sanyal, G., Samanta, D., Nguyen, G. N., & Le, D.-N., Admission control algorithm based-on effective bandwidth in V2I communication. *IET Communications*, 10. https://doi.org/10.1049/iet-com.2017.0825. Online ISSN 1751-8636.
35. Hall, M., Frank, E., Holmes, G., Pfahringer, B., Reutemann, P., & Witten, I. H. (2009). The WEKA data mining software: An update. *SIGKDD Explorations Newsletter, 11*(1):10–18.
36. Settles, B. (2009). *Active learning literature survey*. Computer Sciences Technical Report 1648, University of Wisconsin-Madison.
37. Zhang, C., Niu, F., Re, C., & Shavlik, J. W. (2012). Big data versus the crowd: Looking for relationships in all the right places. In *ACL*.
38. Bubeck, S., & Cesa-Bianchi, N. (2012). Regret analysis of stochastic and nonstochastic multi-armed bandit problems. *Machine Learning, 5*(1), 1–122.

Visualization of COVID-19 Pandemic: An Analysis Through Machine Intelligent Technique Toward Big Data Paradigm

Manash Sarkar, Saptarshi Gupta, Bhavya Gaur, and Valentina E. Balas

Abstract The word pandemic is scary to whole world. It is a pestilence of an irresistible ailment that has spread over an enormous district, for example, different mainland's or around the world, influencing a generous number of individuals. An across the board endemic illness with a steady number of contaminated individuals is definitely not a pandemic. Far reaching endemic maladies with a steady number of contaminated individuals. The people are infected from various part of the world. The number of confirmed cases of COVID-19 in increasing abruptly day by day. The world economy is hampered due to this pandemic. To improve the world economy, it is required to adapt an optimistic decision for during this pandemic as well as for post-pandemic. In this research paper, a real-time data set from World Health Organization (WHO) is collected and analyzed. The nature of the data set is as the concept of big data. A statistical analysis is performed on the data set and produce the results. The data for India is only considered for analysis. The number of death cases and number of COVID-19 confirmed cases in India during the period of 15 weeks from 28th February 2020 to 7th June 2020. Finally, the analysis reported that in India rate of death cases is less than rate of cure cases.

Keywords Pandemic · COVID-19 · Big data · Sampling distribution · Data visualization

M. Sarkar (✉)
Computer Science & Engineering, SRM Institute of Science & Technology, Delhi-NCR, India
e-mail: manash.sarkar26@gmail.com

S. Gupta
Electronics and Communication Engineering, SRM Institute of Science & Technology, Delhi-NCR, India

B. Gaur
Business Technology Analyst, ZS Associates, Pune, India

V. E. Balas
Automatics and Applied Informatics, Aurel Vlaicu University of Arad, Arad, Romania

R. Kumar et al. (eds.), *Multimedia Technologies in the Internet of Things Environment*, Studies in Big Data 79, https://doi.org/10.1007/978-981-15-7965-3_8

1 Introduction

Massive quantity of data can be used for various profits in daily life, business, health care, banking, retail, etc. Proper analysis of big data can be generate revenue as well as human health can also be monitored and proper treatment can be provide to the patients. The patients data can be collected from hospital patient personal file, doctor prescription, Generic databases, wearable devices, IoT healthcare device database, pathology test report, medical imaging, etc. [1, 2]. If massive outbreak of any disease takes place in any county or a particular place that data can be monitored from the government official website. Proper analysis of those data can be led to find some prominent findings and decision. Big data can be utilized to anticipate if there is a pestilence epidemic, remedial diseases, understanding a patient's medical record, keep away from preventable fatalities, etc. Furthermore, patient treatment will be simpler if his health information is collected and the recognition of any severe illness is done in the early hours [3, 4]. A pandemic is an epidemic that is spread various nations. Not all infectious disease terms are made equivalent; however, regularly they are erroneously utilized conversely. The qualification between the words "pandemic," "epidemic," and "endemic" is normally obscured, even by clinical specialists. This is on the grounds that the meaning of each term is liquid and changes as infections become pretty much pervasive after some time. While conversational utilization of these words probably would not require exact definitions, realizing the thing that matters is essential to assist you with bettering comprehend general wellbeing news and fitting general wellbeing reactions. The coronavirus pandemic is probably going to keep going up to two years and would not be controlled until around 66% of the total population is invulnerable, a gathering of specialists said in a report.

In India, the COVID-19 pandemic is a part of the overall pandemic of coronavirus infection 2019 (COVID-19) brought about by serious intense respiratory condition coronavirus 2 (SARS-CoV-2). The first instance of COVID-19 in quite a while, which started from China, was accounted for on 30 January 2020. Starting at 10 June 2020, the MoH and FW has affirmed an aggregate of 276,583 cases, 135,206 recuperations (counting 1 movement) and 7745 passing in the nation. India right now has the biggest number of affirmed cases in Asia, with the quantity of absolute affirmed cases breaking the 100,000 blemish on 19 May and 200,000 on 3 June. India's case casualty rate is generally lower at 2.80%, against the worldwide 6.13%, starting at 3 June. Six urban communities represent around half of every single revealed case in the nation—Mumbai, Delhi, Ahmedabad, Chennai, Pune, and Kolkata. Starting at 24 May 2020, Lakshadweep is the main district which has not revealed a case.

The United Nations (UN) and the World Health Organization (WHO) have applauded India's reaction to the pandemic as "far reaching and hearty," naming the Lockdown limitations as forceful yet essential for containing the spread and building important medicinal services foundation. The current year's World Health Statistics report clarifies that the worldwide endeavors in ongoing decades have been paying off. Taking a gander at the most forward-thinking information organization have on a portion of these imperative Sustainable Development Goals (SDG) pointers, it uncovers wellbeing patterns across Member States, areas, and the whole world. The information shows that people are proceeding to gain enormously promising ground from numerous points of view—yet in addition that they need progress in different manners. Disparity endures, with certain districts despite everything falling behind. They should keep on cooperating to stay concentrated on our objectives. Absence of accessible information called information blind spots, and stay a dire test yet in addition an incredible chance known as rarity of information assortment. Since any place people can quantify, they can gain ground. In light of its capacity to spread from individuals who do not seem, by all accounts, to be sick, the infection might be more enthusiastically to control than flu, the reason for most pandemics in late history, as indicated by the report from the Center for Infectious Disease Research and Policy at the University of Minnesota. Individuals may really be at their generally irresistible before side effects show up, as per the report.

The future of the world of post-pandemic depends on the effect of the presents. To increase the dross domestic product (GDP) rate, an intelligent analysis is required. Due to the COVID-19, people are infected and some worst cases death also. Therefore, the data set which is handled by WHO is increasing abruptly. The nature of the data set is increasing volume, velocity, variety, value, and veracity which satisfied the concept of big data. In this research paper, the data from WHO is analyzed and prepared a statistical report on Indian situation during this pandemic. The data is considered from 28th February 2020 to 7th June 2020. Uniform distribution and sample distribution are applied to determine the degree of closeness for a given sample with COVID-19 confirm cases. The rate of spreading of COVID-19 in India with respect to rest of the world is also compared.

The rest of the paper is defined as follows: details of literature survey is shown in Sect. 2 as a related works followed by fundamental of big data in Sect. 3. Pandemic in perspectives of big data is described in Sect. 4. Mathematical analytics is explained in Sect. 5 followed by data preparation in Sect. 6. Results and simulations are presented in Sect. 7 and discussion in Sect. 8. Finally, conclusion and further scope of research are discussed in Sect. 9.

2 Related Works

As per the history, last few century world is suffered from various pandemic. In 1920, due to influenza pandemic huge number of deaths and immunoprotection cases occurred. In 1817 to 1824, first Asiatic cholera pandemic or Asiatic cholera spread south Asia and South-East Asia. The world is experiencing different new outburst in past years, e.g., Zika virus (Identified in 1947) [5], SARS-CoV—severe acute respiratory syndrome coronavirus (Identified in 2003) [6], Ebola virus (Discovered in 1976) [7], and now novel coronavirus disease 2019 [COVID-19]. The consequences of the 2014 Ebola upheaval in West Africa, general wellbeing experts around the world, were approached to plan for the following pestilence. The biggest flare-up of its sort, the Ebola infection caused 28,600 cases and 11,325 passing's, as per the

US Habitats for Disease Control and Prevention [8, 9]. Serving not just as an exercise for the future, the flare-up additionally featured the significance of enormous information in these sorts of crisis circumstances. Chen et al. [10] depicted for worldwide and national scourges; large information has progressed to a spot where it can give crisis reaction groups ongoing devices and innovation that can possibly screen, contain, and even stop the spread of infection. Conventional wearable devices have various drawbacks, such as un-comfortable for long-term wearing, and insufficient accuracy, etc. Thus, health monitoring through traditional wearable devices is hard to be sustainable. In order to obtain and manage healthcare big data by sustainable health nursing, the system design “smart clothing,” enabling unobtrusive collection of various physiological indicators of human body. To offer persistent cleverness for smart clothing erection, mobile healthcare cloud stand is constructed by the usage of mobile internet, cloud computing, and big data analytics. This paper announces design facts, key tools, and applied implementation methods of smart dress system. Typical claims powered by smart clothing and big data clouds are presented, such as medical backup response, emotion care, disease diagnosis, and real-time tangible interaction. Wang et al. [11] in their research, they estimated the lung cancer using double dispensation system. The image dispensation system was familiarized into double for early prediction. The challenge in this progression was recognition of tiny nodes which comprehends early cancer detection. The unstipulated knobs in lungs can be spotted using ridge recognition algorithm. Kelvin et al. [12] discussed the smoking behavior of the user. The e-cigarette has a small electrical resistance coiled wire in 1.5 Ω which is connected to the positive and negative poles of the device. When the button of e-cigarette is pressed, the resistance coil can be connected with electrical supply under the immersion of some “E-liquid,” the coil heats up and transform the E liquid to vapor, which can be inhaled by the smokers. It monitors the smoking behavior of the user in order to prevent the patient from cancer.

3 Fundamental of Big Data

Big data is a domain in which it deals with the extraction of information from huge amount of data or large data set and analyze the large data components. Nowadays due to huge and large amount of data, the conventional data processing software is not very much suitable for the data analysis and information extractions.

As per Gartner [13] the big data can be clarified as follows:

Big data is high-volume, high-velocity and/or high-variety information assets that demand cost-effective, innovative forms of information processing that enable enhanced insight, decision making, and process automation.

Data complexity increases day by day and this is related to the combinations of five “V” (Characteristics of big data [14, 15]. They are *volume, velocity, variety, value, and veracity*.

- ***Volume***: Huge amounts of data, the data sets size may be near about 1024^7 bytes.

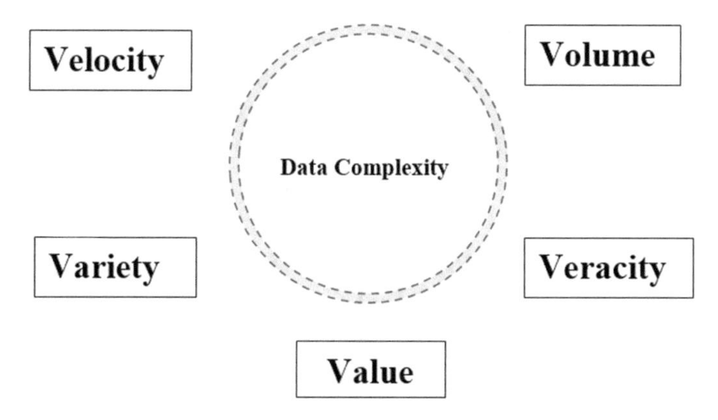

Fig. 1 Properties of big data

- ***Velocity***: High speed data from live or real-time source.
- ***Variety***: Data come from dissimilar sources. Data may be in different formats, e.g., structured (database table), semi-structured (Extensible Markup Language—XML data), unstructured (text, video, image and many more) etc.
- ***Value***: Information in itself is of no utilization or significance; however, it should be changed over into something important to separate Information.
- ***Veracity***: Veracity essentially implies the level of dependability that the information brings to the table. Since a significant piece of the information is unstructured and superfluous, big data needs to locate a substitute method to sift them or to decipher them through as the information is urgent in business improvements.

The big data term is not only indicated the large amount of data also it indicates the data arrives in fast speed having complex format from variety of sources. Basic properties of big data are graphically represented in Fig. 1.

3.1 Types of Data

Web Data, Text Data, Weather Data, Class Room Data, Time and Location Data, Email Data, Social Network Data, Sensor Data, Data from different Industries and Organizations, etc. (Fig. 2).

3.2 Big Data Ecosystem Components

The excursion to big data is to comprehend the levels and layers of abstraction, and the parts around the equivalent. Some basic segments of big data investigative stacks and their combination with one another. The proviso here is that, in a large

Fig. 2 Various big data sources [16]

portion of the cases, HDFS/Hadoop structures the center of the greater part of the big-data-driven applications, however that is not a generalized rule or guideline.

3.2.1 Technologies for Capturing, Storing, and Accessing Big Data

Big data's processing should be possible in four layers as appeared in Fig. 3. The primary testing task is to gather colossal measure of data from various sources in various arrangement. As the data is unstructured, it gets hard for conventional database the executives framework to separate information out of it yet huge data take care of this issue since it might help in extricating the information from organized, semi-organized, and unstructured data.

Initial step is to gather the data that is assembled from various sources and afterward store this gathered data in some common spot. To give distributed file system (HDFS) for conveyed capacity and adaptation to non-critical failure, Apache Hadoop is normally being utilized nowadays. MapReduce is a programming model utilized in Hadoop for handling enormous measure of data rapidly. Machine learning (ML) algorithm can be applied for accomplishing fast examination on input data and make the data which can be utilized for delivering data in processing layer.

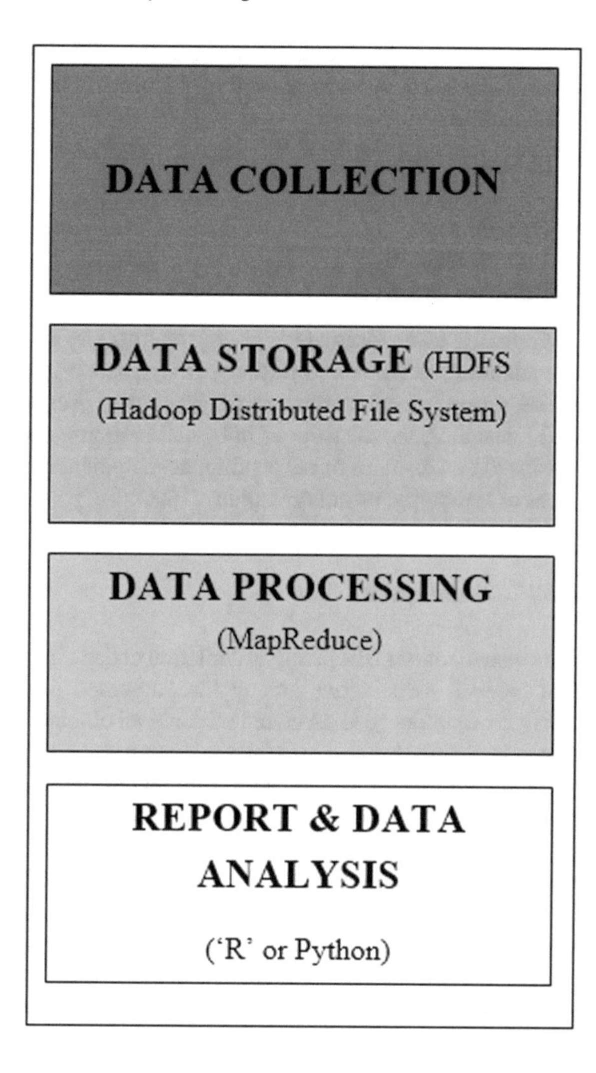

Fig. 3 Big data processing steps block diagram [17]

3.2.2 Analytical Techniques

Data analysis (DA) is the way toward analyzing data set (inside the type of text, sound and video), and reaching determinations about the data they contain, all the more usually through explicit frameworks, programming, and strategies. Data analytics advancements are utilized on a modern scale, across business ventures, as they empower associations to make determined, intelligent decision for business.

Big data analysis mixes conventional measurable data analysis approaches with computational ones. Statistical sampling from a universal set is perfect when the whole data set is accessible, and this condition is commonplace of conventional bunch preparing situations. Be that as it may, big data can move clump handling to real-time preparing because of the need to understand spilling information. With gushing information, the data set aggregates after some time, and the information is time-requested. A large portion of the broadly utilized analyzing techniques are listed as follows:

- Statistical methods
- Machine learning
- Data mining
- Data warehouse
- Database querying.

Globally, enterprises are bridling the intensity of different various information investigation methods and utilizing it to reshape their business models. As innovation grows, new examination programming rises, and as the Internet of Things (IoT) develops, the measure of information increments. Enormous information has advanced as a result of our expanding development and association, and with it, new types of removing, or rather "mining," data.

3.2.3 Visualization

Data visualization is the practical depiction of data. It incorporates conveying pictures that pass on associations among the addressed data to watchers of the photos. This correspondence is practiced utilizing an efficient mapping between reasonable engravings and data regards really taking shape of the portrayal. This mapping sets up how data regards will be addressed ostensibly, choosing how and how much a property of a reasonable engraving, for instance, size or concealing, will change to reflect changes in the estimation of a datum. To grant information indisputably and capably, data discernment uses genuine structures, plots, information delineations, and distinctive devices. Numerical data may be encoded using spots, lines, or bars, to apparently bestow a quantitative message. Convincing portrayal helps customers with separating and reason about data and evidence. It makes complex data progressively accessible, legitimate, and usable. After analysis of the data, graph needs to be plotted to see the results and data visualization is necessary. "R" language can be used for data visualization.

3.3 Applications of Big Data

In today's world, the data is wealth. Based on the previous data analysis, the future plan would be accepted by any organizations as well as individuals also. Today, people are very much flexible to access their information through Internet. In the concept of Internet of Things (IoT)-based smart cities, the life is totally controlled by the internet. Due to precious time in human life and their precious data to keep in a secure place, people are going to store their data in various cloud platform. Therefore, the volume of the data and their velocity, variety are also changes very rapidly. In every fields like social networking, health care system, online shopping, online traveling system, academic system, finance system, agriculture, concept of big data is explored. There are some application of big data [18] are discusses as follows:

- ***Segmentation and prediction***: According to the customers behavior, segmentation can be done and different predictions can be formed for growth of the business.
- ***Recommender systems and marketing***: Various recommendation schemes are already available in existing world but that may be placed in a proper place. Nowadays, online recommendation system become very famous, and by using this recommendation system, business industries can target the appropriate customers or audience that can lead to target marketing.
- ***Sentiment analysis***: This is one of the new trend in data analysis. This kind of examination offer data on what the market is stating, thinking & feeling about an association. Mostly the sentiment of a customer or individuals is analyzed by the data taken from social media.
- ***Operational analytics***: Without the human involvement, automatic decision can be taken in consumer service in different industries so that quick and fair decision can be made for all the consumer.
- ***Prediction for optimal decision***: This is the one of most important domains where the application of big data is fully explored. Analysis of previous data set and trace out an optimal decision for the near future. Most of the organizations even also individuals are considered for this services. In the field of marketing, production, banking sector, health care, economic are the most useful domain for the application of big data.

In this paper, a data set of pandemic of current scenario is considered and analyzed to predict the infection of COVID-19 for the people of India. This is one kind of big data application. The data value of the data set is changed abruptly with variety of values.

4 Pandemic: Perspectives of Big Data

The episode of COVID-19 has seriously influenced all parts of human life, forcefully affected the worldwide economy, and has put an enormous strain on the worldwide wellbeing framework. In an offer to contain the pandemic, organizations and pioneers at all levels are utilizing huge information and investigation devices, among different arrangements, to bring down the effect of the infection. They are utilizing AI-enabled framework, analyzing, and visualization tools for big data to conceive the eventual fate of the pandemic, track the continuous spread of the infection, discover cures against COVID-19.

Various associations likewise have begun to utilize huge information examination advances to speed up sedate improvement against COVID-19 and better see how the invulnerable framework vanquishes the infection. Pharmaceutical organizations GlaxoSmithKline (GSK) and Vir Biotechnology, for example, joint powers to progress coronavirus treatment improvement utilizing artificial intelligence.

Abruptly spreading of COVID-19s and effect have caused individuals in the whole world to feel vulnerable, terrified, and baffled. Be that as it may, the once in a while news and disclosures about the fix and utilization of current innovation sponsored by data makes them to feel safe and expectation in the terrible "pandemic" time.

While expressions of social distancing, lockdowns, and straightening the bend are turning out to be basic aphorisms in a typical man's life, big data is demonstrating clever in considering the measurements about populace developments across districts, check open consistence in following the lockdown and wellbeing conventions, size and recurrence of individuals with higher paces of temperature from the information by temperature scanners, all of which help in foreseeing how the curve with develop or decrease in the coming weeks.

In order to establish the decision on data-driven and convey powerful administrations, organizations, policymakers, government substances, and others over the globe are understanding the operational preferences of utilizing enormous information examination. Moreover, intensified speculation from government and intelligence (G&I) and areas of social health care to deal with the pandemic will drive the market in the post-pandemic world. The capacity of big data to gather, store volumes of information, speed to permit continuous information stream for quicker preparing, bolster an assortment of data form: unstructured or structured, numeric-emblematic, from various sources and inconstancy in taking care of them according to request loads with higher veracity makes it an extraordinary favorable position when utilized alongside artificial intelligence, machine learning, IoT applications, etc. Methods of big data analytics are appropriate for tracing and managing the spread of COVID-19 around the world.

5 Mathematical Analytics

In this research, data analysis is done where the volume of the data is enormous and the pattern of the data is unstructured and there is a veracity of the information.

The data set of COVID-19 is considered as a test bed. The properties of the data set are nothing but big data. Various state of India has been affected by COVID-19, but number of infected persons are vary from state to state.

Assuming total population of India is considered as a formal with coordinate with system $\{x_1, x_2, \ldots, x_n\}$ Let consider the density function $\delta(x_1, x_2, \ldots, x_n)$ which is the limit for the ratio of the number of infected in different state. The population of a small region is Δv round the point $\{x_1, x_2, \ldots, x_n\}$ to Δv. The total population set will be equilibrium if the infected persons are distributed as closet as possible to equidistribution over the total population with kinematic restriction.

The sample infected of position $\{x_1, x_2, \ldots, x_n\}$ has $E(x_1, x_2, \ldots, x_m)$ the mean of infected per each sample is given constant as

$$\int ES\,dv = d \tag{1}$$

where dv indicates the unit volume.

At uniform distribution, determine δ. The degree of closeness for a given sample distribution to uniform distribution. As the pattern of data as a big data, high degree of closeness is required. The entropy of the sample is defined by Eq. (2).

$$-\int S\ \log s\, dv \tag{2}$$

The solution of Eq. (2) as maximizing is subjected to Eq. (1)

The optimum choice of samples S for which $-\int S \log s\, dv$ will be maximize depends on $\int ES\,dv = d$ and is defined as

$$S = \mathrm{e}^{\lambda E+\mu} = \alpha \mathrm{e}^{\lambda E} \tag{3}$$

where α is selected such that $\int Sdv = 1$

As per the inequality

$$\int S \log \frac{s}{q} \geq 0 \tag{4}$$

For every application of any two alternative density is defined as

$$-\int S \log s\, dv \leq -p \log\, q\, dv = -\int Z(\lambda E + \mu)dv = -(\lambda d + \mu) \tag{5}$$

If Eq. (5) is applied for Eq. (1), then the fact will be integral of S is unity.

The data samples for infected persons are considered for different state of India. The rate of infection for different state would be different. Sampling distribution is deployed to transform the rate of infection at various level. For linear transformation $Y = BS, \forall S \rightarrow Y.$

If B is considered a non-singular matrix then

$$\frac{DY}{DS} = |B| \quad \text{as a positive sign} \tag{6}$$

The interrelated differential elements are defined by the relational equation in Eq. (7).

$$dy_1\, dy_2 \ldots dy_n = |B|\, ds_1\, ds_2 \ldots ds_n \tag{7}$$

From Eq. (6)

$$dY = |B|dS \quad \forall\, B \quad \text{is an orthogonal matrix and } |B = 1| \tag{8}$$

$Y = BS$ which transforms into a quadratic form

$$S^I S \to Y^I Y$$
$$(S - \mu)'(S - \mu) \to (Y - \eta)'(Y - \eta) \quad \text{where, } \eta = B\mu \tag{9}$$

After determine the dissimilarity of the infected and suspected patients, distance from threshold value is evaluated by applying partitioning matrix.

$$B = \begin{pmatrix} B_1 \\ \cdot \\ \cdot \\ \cdot \\ B_k \end{pmatrix} \quad \text{where,} \quad B_i = n_i \times n \quad \text{and} \quad \sum n_i = n \tag{10}$$

The partitioning matrix is partitioned into kth sub-matrices. All sub-matrices are orthogonal to each other, but not be orthogonal themselves.

6 Data Preparation

A real data set is collected from World Health Organization (WHO) [19]. Data is collected on 10 June 2020 from official website of WHO. Only nineteen individual data records are collected. Each record contains two attributes like number of confirmed cases and number of deaths for seven days intervals. Each record contains the same details of three different regions in the world; India, South-East Asia Region, and world. In this research, the data is considered from 28th February to 7th June 2020; total 15 weeks in intervals. In this data set, total number of confirmed cases from each states in India are not mentioned. As per the requirement of the proposed methodology, data set is prepared. The modified data set is shown in Table 1.

For further classification and analyzing, different derived data sets are generated. Table 2 is derived from Table 1 and it shows number of increases for confirm cases in India during every seven days intervals. Table 2 contains only the records of India for the same time intervals.

Table 2 is a derived from Table 1. The data set, collected from WHO, is considered the death cases only caused by COVID-19 (Table 3).

7 Results and Simulations

The results are evaluated and simulated with MATLAB 2018a and R statistic tool. Values from the tables are used to evaluate the results. In this research, main focus is the record of COVID 19 confirmed for India. The data is taken only for 15 weeks

Table 1 Data set World Health Organization

S. No.	Date	No. of State in India	Confirmed (India)	Death (India)	Confirmed (South-East Asia Region)	Death (South-East Asia Region)	Confirmed (World)	Death (World)
1	28-Feb-2020	1 (Kerala)	3	0	–	–	–	–
2	09-Mar-2020	11	44	0	–	–	–	–
3	14-Mar-2020	13	84	2	–	–	132,758	4955
4	22-Mar-2020	23	360	7	979	38	266,073	11,184
5	28-Mar-2020	27	909	19	2536	79	462,684	20,834
6	05-Apr-2020	29	3577	83	6528	267	1,051,635	56,985
7	12-Apr-2020	31	8447	273	14,161	617	1,610,909	99,690
8	19-Apr-2020	32	16,116	519	27,319	1185	2,241,359	152,551
9	26-Apr-2020	32	26,917	826	41,073	1658	2,719,897	187,705
10	03-May-2020	32	40,263	1306	60,490	2256	3,267,184	229,971
11	10-May-2020	32	62,939	2109	90,808	3204	3,855,788	265,862
12	17-May-2020	32	90,927	2872	127,995	4201	4,425,485	302,059
13	24-May-2020	32	131,868	3867	182,278	5556	5,103,006	333,401
14	31-May-2020	32	182,143	5164	260,579	7431	5,934,936	367,166
15	07-Jun-2020	32	235,657	6642	336,577	9316	6,663,304	392,802

Table 2 Number of increases of confirmed cases

Days	Number of increases in India
0	0
7	41
14	40
21	276
28	549
35	2668
42	4870
49	7669
56	10,801
63	13,346
70	22,676
77	27,988
84	40,941
91	50,275
98	53,514

Table 3 Increases of death with respect to time

Days	Death cases in India
0	0
7	0
14	2
21	7
28	19
35	83
42	273
49	519
56	826
63	1306
70	2109
77	2872
84	3867
91	5164
98	6642

started from 28th February 2020 to 7th June 2020. During this time period, the number of COVID 19 confirmed cases is gradually increased up to sixth week then suddenly it is changed very abruptly.

In Fig. 4, the nature of increasing through 15 weeks is plotted. At seventh week,

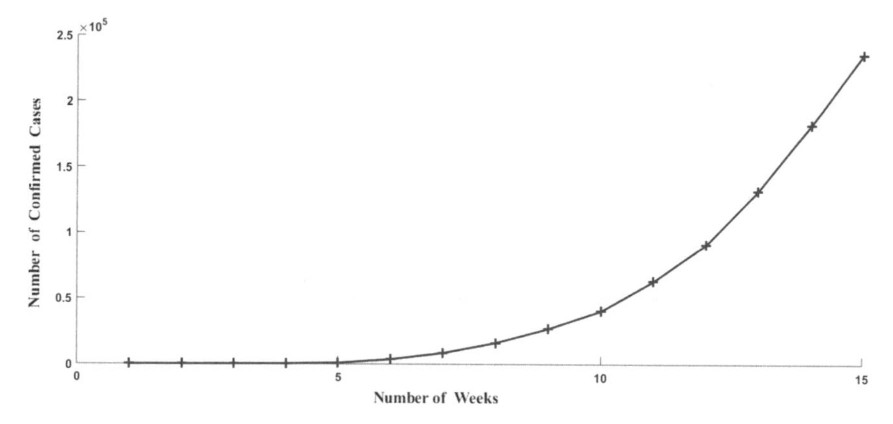

Fig. 4 Number of confirmed cases with number of weeks in India

the graph rapidly increases. After seventh weeks, how the confirmed cases increase and corresponding their rate of increasing also changes based on the number of confirmed cases. In Fig. 5, the rate of increasing is shown with time domain.

After analyzing Fig. 4, it is clear that the rate of increasing of confirmed COVID-19 is abruptly changed at 42 days. As the rate of confirmed cases is increased, contemporary number of deaths due to COVID-19 are also increased. The number of death cases due to COVID-19 is directly depended on the rate of confirm infected cases. Therefore, the number of death cases is also increased as the number of confirmed cases is increased.

Figure 6 represents the visualization of the analytical report of death cases due to COVID-19. The death rate is also abruptly increased from seventh week onwards. How number of death cases and number of confirmed cases are varied with each

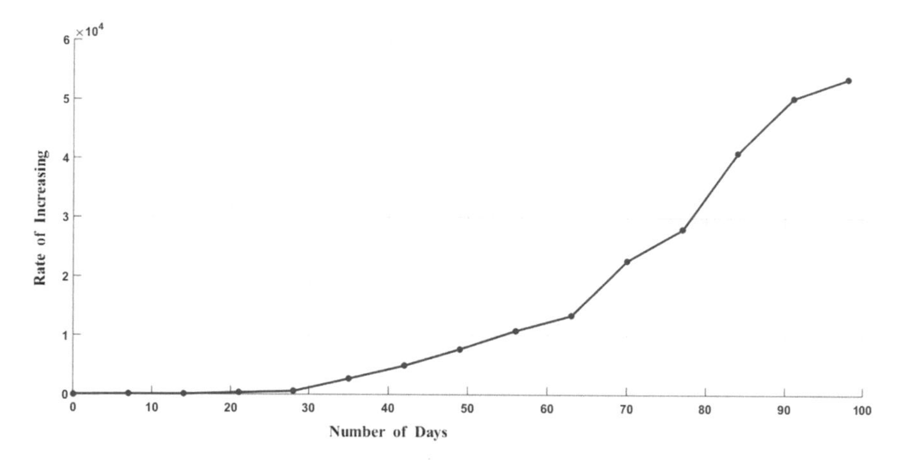

Fig. 5 Rate of increasing of confirmed cases during days in India

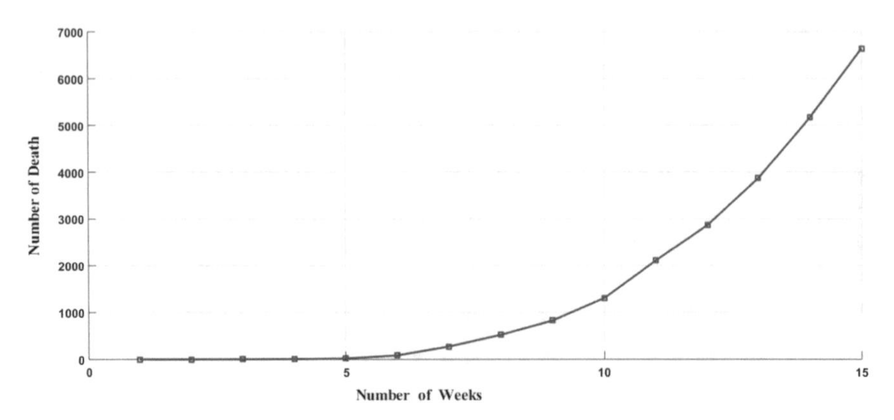

Fig. 6 Number of death cases in weeks intervals in India

other's in a particular time domain, is shown in Fig. 7. In India, after 15 weeks, the total number of death cases are 6642, whereas the number of confirmed cases are 235,657.

After 15th weeks, the rate of death in respect to total confirmed cases is very less, it is 2.81%. In Fig. 7, it is shown that the graph of death cases is very closest and parallel to Y-axis whereas, the graph of confirmed cases is increased extremely. From this research analysis, it is clear that the rate of death is gradually decreases in respect to number of confirmed cases.

In support of this statement, another death report of other organization is shown Fig. 8 which convergences to this proposed research analysis. Report at Fig. 8 is

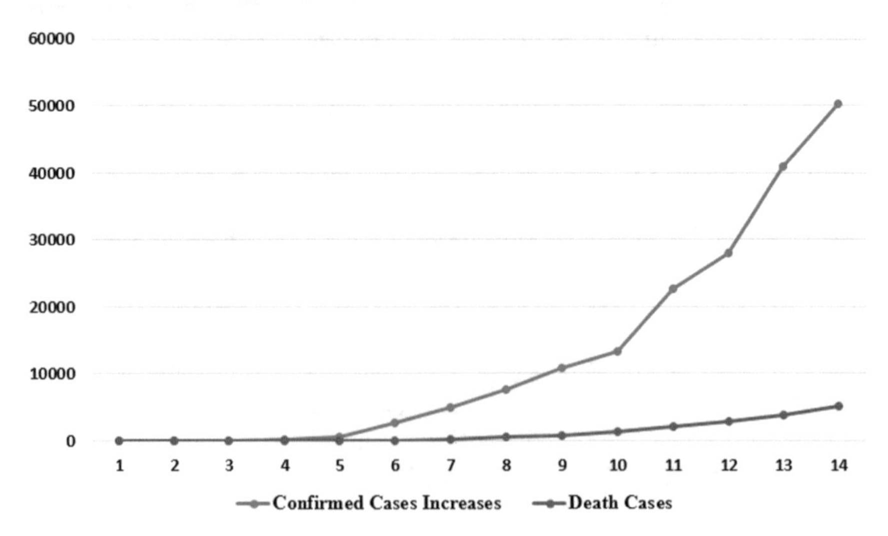

Fig. 7 Death cases versus confirmed cases

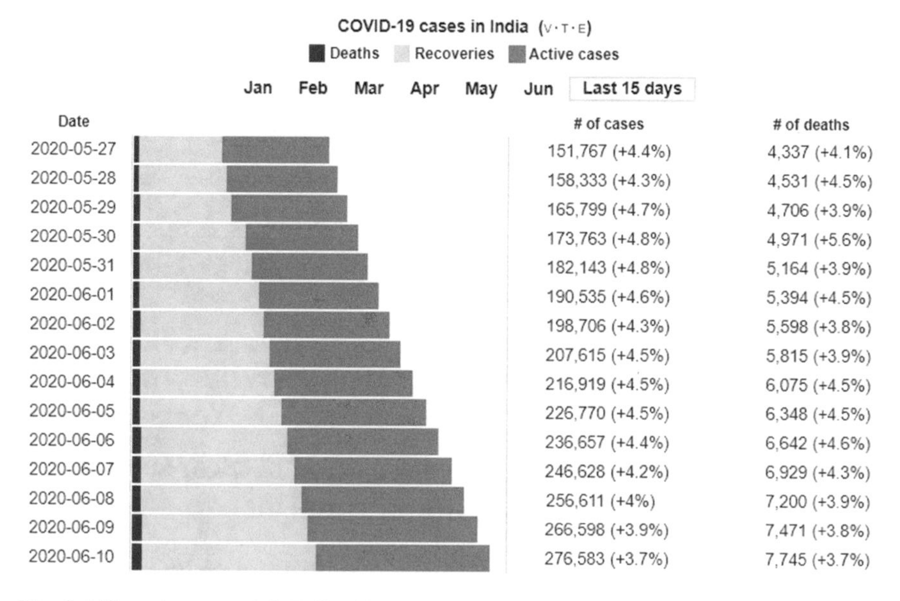

Fig. 8 Fifteen days records in India [20]

taken from 27th May 2020 to 10th June 2020. The report shows that the percentage of death on 27th May 2020 is 4.1% while it is 3.7% on 10th June 2020.

8 Discussion

As per the report of WHO, the data is collected and analyzed through equidistribution over the total population. The main focus of this paper is the analysis for spreading of COVID-19 in India and also compares the rate of increasing with rest of the world. If a clear look into the data of India, then it is found that on 28th February 2020, India has only three confirmed cases of COVID-19, whereas, on 7th June 2020, it has been increased on 235,657. The report of spreading coronavirus in India is very shocking. But if think deeply, then it is found that the number of deaths during this 15 weeks is changed from 0 to 6642. The number 6642 is really mean as a death cases. But if a statistical analysis is explored, then find the mean (μ) value and standard deviation (σ) in both cases.

$$\mu_{\text{Confirmed}} = \frac{1}{n}\sum_{1=1}^{n} \text{Confirm}_i \quad \text{and} \quad \sigma_{\text{Confirmed}} = \sqrt{\frac{\sum (p_i - \mu_{\text{Confirmed}})^2}{n}}$$

where $i = 1, 2, \ldots, n$.

$$\mu_{\text{Death}} = \frac{1}{n}\sum_{1=1}^{n} \text{Death}_i \quad \text{and} \quad \sigma_{\text{Death}} = \sqrt{\frac{\sum (p_i - \mu_{\text{Death}})^2}{n}}$$

where $i = 1, 2,\ldots, n$.

where

$n \rightarrow$ Size of the population

$p_i \rightarrow$ Each value from the population.

In this paper, the value of $n = 15$

$$\mu_{\text{Confirmed}} = 15{,}710.27 \quad \text{and} \quad \sigma_{\text{Confirmed}} = 18{,}988.04$$
$$\mu_{\text{death}} = 1579.26 \quad \text{and} \quad \sigma_{\text{Death}} = 2134.092$$

Now, look into the death rate during this considerable 15 weeks. From Table 1, it is found that on 28th February 2020, in , death cases are zero whereas, on 14th 2020, March, death cases are 2. On 7th June, death cases in India are 6642. Now, evaluate the percentage of death cases in respect to the number of confirmed cases.

In India, the rate of death cases based on the number of confirmed cases during different time period are shown in Table 4. The death rate of India is compared with the death rate of world.

Figure 9 shows a Pie chart to describe the death rate of India with death rate of world due to coronavirus COVID-19 on 7th June 2020. From Table 4, it is cleared that in India, the death cases are not increased so abruptly as the number of confirmed cases. It gives some hopes of better tomorrow.

Table 4 Rate of death cases at various time period

Time duration	Percentage of death cases (%)
On 14th March 2020, percentage of death	1.17
On 7th June 2020, percentage of death	2.81
31st May to 7th June, death rate increases	2.76
Average death rate (14th March to 7th June)	1.64

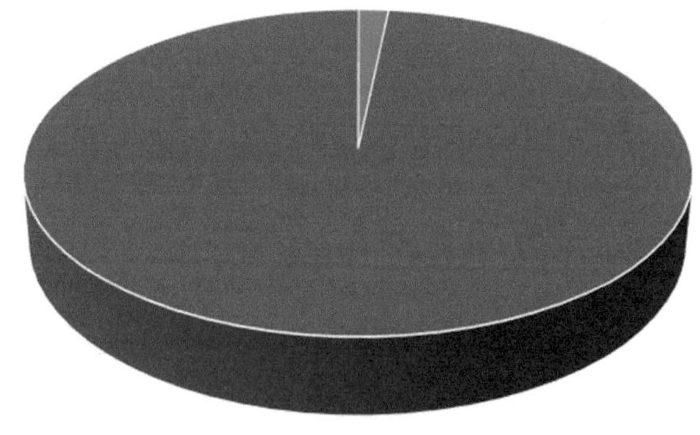

Fig. 9 Percentage of death rate of India and death rate of world

9 Conclusion

The COVID-19 pandemic, otherwise called the coronavirus pandemic, is a continuous pandemic of coronavirus sickness 2019 (COVID-19), brought about by serious intense respiratory disorder coronavirus 2. The infection is principally spread between individuals during close contact, regularly by means of little beads delivered by hacking, wheezing, and talking. The beads generally tumble to the ground or onto surfaces as opposed to going through air over significant distances. The whole world is suffering from COVID-19 corona virus. Number of people is infected abruptly and contemporary death rate is also increased. In this paper, a statistical analysis is performed over the data set collected from WHO. A sample distribution is performed on the data of India from 28th April 2020 to 7th June 2020. During this 15 weeks how the rate of change of COVID-19 confirmed cases and also death cases in India are determined and visualized by simulations outputs. On 14th march 2020, the death rate of India is 1.17% in respect to whole infected people, whereas it is increased up to 2.81% on 7th June 2020. From the analysis report, it is cleared that the rate of increasing death is very less than rate of increasing infected cases. Thus, it implies that the rate of cure is greater than rate of death. The analysis of the pandemic data set as a big data is useful to make an optimistic decision for post-pandemic scenario. Based on this analysis, the economic condition, GDP rate, market analysis will be considered as future research directions.

References

1. Mathawan, R. (2020). Big Data in Healthcare, March 25, 2020. https://techstory.in/big-data-in-healthcare/. Accessed on 12/06/2020.
2. Dash, S., Shakyawar, S. K., Sharma, M., & Kaushik, S. (2019). Big data in healthcare: management, analysis and future prospects. *Journal of Big Data, 6*(54), 1–25.
3. Saranya, P., & Satheeskumar, B. (2016). A survey on feature selection of cancer disease using data mining techniques. *International Journal of Computer Science and Mobile Computing,*

5(5), 713–719.
4. Hamilton, B. (2012). Big Data is the future of healthcare, cognizant white paper, 2012. https://www.cognizant.com/industries-resources/healthcare/Big-Data-is-the-Future-of-Healthcare.pdf. Accessed on 10/06/2020.
5. https://www.who.int/news-room/fact-sheets/detail/zika-virus#:~:text=Zika%20virus%20is%20a%20mosquito,Americas%2C%20Asia%20and%20the%20Pacific. Accessed on 10/06/2020.
6. https://www.who.int/ith/diseases/sars/en/. Accessed on 10/06/2020.
7. https://www.who.int/news-room/fact-sheets/detail/ebola-virus-disease. Accessed on 10/06/2020.
8. https://www.cdc.gov/vhf/ebola/history/2014-2016-outbreak/index.html. Accessed on 10/06/2020.
9. https://onlinedegrees.unr.edu/blog/the-role-of-big-data-in-global-epidemics/. Accessed on 10/06/2020.
10. Chen, M., Ma, Y., Song, J., Lai, C.-F., & Bin, Hu. (2016). Smart clothing: Connecting human with clouds and Big Data for sustainable health monitoring. *Mobile Networks and Applications, 21,* 825–845.
11. Wang, W., & Wu, S. (2006, June 25–28). A study on lung cancer detection by image processing. *International Conference on Communications, Circuits and Systems.*
12. Tsoi, K. K. F., Kuo, Y.-H., & Meng, H. M. (2015). A data capturing platform in the cloud for behavioral analysis among smokers an application platform for public health research. In 2015 IEEE International Congress on Big Data, New York, NY, USA, 27 June–2 July 2015.
13. Beyer, M. A., & Laney, D. (2012). The importance of 'Big Data': A definition. Gartner.
14. O'Leary, D. E. (2013). Artificial intelligence and big data. *IEEE Intelligent Systems, 28,* 96–99.
15. Berman, J. J. (2013). Introduction. In *Principles of Big Data* (pp. xix–xxvi). Morgan Kaufmann, Boston.
16. https://www.ubuntupit.com/best-big-data-applications-in-todays-world/. Accessed on: 23/05/2020.
17. Zikopoulos, P., et al. (2011). *Understanding Big Data: Analytics for enterprise class hadoop and streaming data.* McGraw-Hill Osborne Media.
18. Su, X., Introduction to Big Data, NTNU Learning material is developed for course IINI3012 Big Data.
19. World Health Organization (WHO). https://www.who.int/india/emergencies/coronavirus-disease-(covid-19)/india-situation-report. Accessed on 10/06/2020.
20. https://en.wikipedia.org/wiki/COVID-19_pandemic_in_India. Accessed on 10/06/2020.

Multimedia Security and Privacy on Real-Time Behavioral Monitoring in Machine Learning IoT Application Using Big Data Analytics

R. Ganesh Babu, K. Elangovan, Sudhanshu Maurya, and P. Karthika

Abstract The Internet of Things (IoT) is an advancing idea where physical articles are associated with one another and customer through the Internet so as to share information among contraptions and structures. Under IoT, splendid contraptions are transmitted to screen, follow, and dissect business, individual and social activities. The IoT's unavoidable idea will conceivably be fruitful; however, it additionally uncovers a risk from malware, developer obstruction, diseases, etc. The subsequent security and insurance issues can cause physical mischief and even compromise human lives. This paper targets two zones; from the start, the paper gives an examination of the IoT layered engineering to set up an IoT space with the security challenges/attack. From that point on, the paper proposes a response that can protect the security of mixed media information within an IoMT state (Internet of Multimedia Things) at its observation layer. The proposed security structure would improve the mixed media to operate adequately while at the same time safeguarding the privacy of the transmitted data and protecting people.

Keywords Internet of Things(IoT) · Internet of Multimedia Things(IoMT) · Multimedia Security · Machine Learning · Big Data Analytics

---

R. Ganesh Babu (✉)
Department of Electronics and Communication Engineering, SRM TRP Engineering College, Tiruchirappalli, Tamil Nadu, India
e-mail: ganeshbaburajendran@gmail.com

K. Elangovan
Department of Electronics and Communication Engineering, Sriram Engineering College, Chennai, Tamil Nadu, India
e-mail: elangocss@gmail.com

S. Maurya
School of Computing, Graphic Era Hill University, Dehradun, Uttarakhand, India
e-mail: dr.sm0302@gmail.com

P. Karthika
Department of Computer Applications, Kalasalingam Academy of Research and Education, Srivilliputhur, Tamil Nadu, India
e-mail: karthikasivamr@gmail.com

R. Kumar et al. (eds.), *Multimedia Technologies in the Internet of Things Environment*, Studies in Big Data 79, https://doi.org/10.1007/978-981-15-7965-3_9

1 Introduction

The ever-expanding Internet-offered administrations and applications have dramatically broadened the worldwide spectrum between organizes. As of now, more 9 billion machine devices are attached to the Internet, attracting over 3.5 billion. Individuals approximately the world of communication (messages, informal organizations, visit rooms, places, conversations, etc.), recreation and distraction, sharing the information (instruction, land details, reference books, etc.), among others. Continuous progress in minimal effort structuring low-scale gadgets, strengthened by developments as Micro-Electro-Mechanical Systems (MEMS), has projected a huge flood of Internet-enabled gadgets.

A risky growth in the quantity of gadgets is gauge over the decade that follows [1]. Therefore, given modern devices, such as personal computers, PCs, mobiles etc., the actual substance or things approximately us will be able to talk to every other [2]. Brilliant things prepared with the ability to watch or theoretically communicate with objective condition and the capability to talk to different things extends the Internet.

Existing articles on IoT characterize it with regard to the identification and impelling capabilities of devices, systems management and cloud developments, and market opportunities; they also find the barriers to institutionalization and assess its broader implications for society [3]. Be that as it may, these exam contemplates do not think alongside other scalar knowledge about the necessities/challenges posed by mixed media gadgets or the transportation of interactive media traffic over network. Innate sight and sound data attributes place specific limitations on IoT structure, despite. The difficulties compelled by numerous heterogeneous devices that are a piece of IoT. In order to meet the criteria of quality of service (QoS), the network individuality distinct by start-to-finish postponement, jitter and error rates, among others, must be managed to ensure decent media content transmission.

Because of the immense enthusiasm for the creation and use of media-based applications and governments, there have been significant improvements in sight and sound rush hour gridlock around the world. Continuous sight and sound technologies, administrations and structures, for example, video conference, remote video-on-request, telepresence, consistent substance transport and online gaming, have prompted the exponential development of Internet blended media traffic. The current equilibrium among non-sight and sound information traffic and media traffic is currently moving specifically as far as video content is concerned, toward an increase of mixed media content. Late inquiries into trends and worldwide Internet traffic figures [4] suggested an optimistic sight boost and a sound rush hour gridlock system in the next five years. Right now, an activity [5] has been conducted to find the trends of applications for visual network administration in the worldwide Internet IP traffic. Depicts single of the key discoveries of this investigation, in which it is evident that sight and sound (video) traffic can essentially overpower Internet IP traffic. This wonder is also relied on to explain quickly the speech of applications on the Internet of Things.

Sight and sound content, such as sound, video, etc., obtained from the physical condition have specific attributes in contrast to the scalar information produced by common IoT gadgets. Given what could be expected, the gadgets for sight and sound require higher handling and processing assets to process the interactive media data gained. In comparison with the standard scalar information traffic in IoT, sight and sound transmission are also more willing to switch speed. Presenting mixed media objects cultivates a wide array of applications in industry as well as military spaces. Some models are: constant sight and sound-based security/monitoring systems in genius homes, remote patients tested in successful clinics with immersive Web-based telemedicine benefits, cautious mixed media. The observation systems sent in clever urban areas, the board updated transport using shrewd camcorders, remote mixed media-based environmental system testing, and so on. Nonetheless, the extension of IoT systems of sight and sound gadgets and substance is not straightforward and involves the introduction of additional functionalities and the correction of existing ones, leading to a specific sub-set of IoT, to which we refer with 'Internet of Multimedia Things' (IoMT).

The state of being information acquired by customary IoT remote sensor devices may incorporate estimations of light, temperature, pressure, etc., and the points of interest of their conditions/state, for example, water level in a water-allocator, battery status, or shortcoming that uncovers perceptive assistance. The idea of this detected data is occasional and requires less computational and memory properties. Thus, such applications require the recognition contraption to plan power clearly and lower data levels. Not at all like what could be normal, the intuitive media information in IoMT is bulky in nature and explicitly for higher arranging and memory advantages for consistent correspondence is required. The making sure about and correspondence of sight and sound by existing IoT gadgets is not practical right now. In IoMT, the transmission of blended media data ought to be inside the constraints of QoS details (for example, delay, jitter) that submit higher exchange speed and proficient specialized instruments. The RPL controlling show in the current IoT correspondence stack is versatile and adaptable as per the application requirements to work in an essentialness agreeable way.

With the aid of sensors and actuators, information and triggering ability are embedded in the gadgets separately in an IoT-based environment. A cloud also empowers the ability to create, sustain, and run different administrations through the provision of flexible figure and capability properties. It helps the clients to screen and monitor the devices from anywhere and whenever they want. Existing cloud servers conveyed for managing IoT boost scaler information that is normally occasional in nature [6]. Then again, in fact the sight and sound knowledge is unceasing, and demands high capacity for planning. In addition, interactive media information requires a high power and capacity limit, and mixed media information handling is particularly boggling and testing in support of ongoing video spilling. There has been a trend for mixed media communication over remote systems in various earlier reviews, but with a limited degree that is an explicit gadget, explicit application, explicit content [7], among others. Nonetheless, none of these exams center on IoMT's system development, which involves media gadget heterogeneity. Data transmission challenges of communication, complicated sight and sound planning, cloud management, and numerous issues. In this way, a completely different engineering for IoMT paradigm needs to be developed that spotlight on these issues as a whole. The purpose of this document is to clarify IoMT's viewpoint vision stimulated by IoT's ideas. In accumulation, the IoMT values are completely broken down and contrasted, and the existing relevant frameworks.

Instead of thinking about a exacting use of an IoMT framework, we current a possible IoMT structural design that isolates the procedure of an IoMT organization into four unmistakable stages; (i) mixed media identification, (ii) description and address ability, (iii) mindful cloud interactive media, and (iv) multi-specialist frameworks. In addition, an analysis of previously existing advances is conducted which gives the recognition of IoMT-based frameworks a union and direction by allowing for various needs and difficulties as fine as the attainability of presented IoMT responses. This paper's principal responsibilities can be summarized as follows:

- We may learn the key manuscript showing the perception and visualization of the IoMT, the implications of which are explored with the aid of specific cases.
- Once compared with the existing sight and sound frameworks, IoMT's unmistakable structural design and quality are thoroughly examined.
- The requirements and preconditions set out in the IoMT Systems are differentiated and addressed.
- Conventions for communications intended for IoT are debated and their practicality for IoMT dissected.

2 IoMT Visualization

In the degree of that job and essential highlights of IoMT are all the more likely to describe, we begin to break down the present qualities of two systems that are now used to express benefits in large interactive IoT media request: the wireless multimedia organization. These are utilized to screen the earth and anywhere the detecting gadgets can be blamed for manipulating the receiving procedure and the Wireless Multimedia Sensor Networks (WMSN), where the sight and sound gadgets have no (or restricted) input capability. WMS has been carried out to provide forms of assistance and services in a few areas, e.g., diagnosis, transport, telemedicine. The WMS structure is for the most part the one delineated, where the information is sent to interactive media outlets. To the Internet through a server, where consumers and managers plan the sight and sound substance and put it away for either synchronous or offbeat access. In case we need a device for remote sight and sound detection and testing. The interactive media devices can be camera hubs, gathering mixed media data from the planet and announcing the remote specific breakthrough back to the control community.

Numerous tasks have been completed, including transit reviews and executive systems expressed in the metropolis of Irving, the institution of higher education of Minnesota [8] and the institution of higher education of North Texas among others.

In several remote video recognition organization, the camera hubs can announce the sight and sound substance to a organize group or to a cloud attendant [9]. The control culture provides criticism to change the state of the camera (switching on or off) or to adjust the location of the camera to change the perspective on suspense. The mixed media gadgets in these operations are the shut-circuit TV (CCTV) cameras with the production capability of fixed or tilt-zoom dish (PTZ) [10], and specific remote advances, for example, IEEE 802.11 or IEEE 802.16 are exploit.

Notwithstanding in Fig. 1 the achievement of traditional remote sight and sound frameworks, notable constraining variables are present which restrict the universal adaptation of these frameworks. Right off the bat, the scope of these structures is strictly restricted to the organizational situation according to a defined design with prohibitive flexibility, pre-characterized collection of sights and sound gadgets that approximately pre-characterized locate of functionalities. Also, the gadgets of sight and sound are generally regulated by the principal source of vitality. There is no constraint on the exploitation of vitality along these lines, with the intention that the structures communicated are not efficient of vitality. Thirdly, the mixed media gadgets with comparative correspondence stacks are not intended to speak to other machine gadgets performing arts various tasks. Consider, for instance, a situation in which cameras are expected to create recording when a exacting sign is obtained from an objective finder sensor. In that, initial the finder appreciation is throwing to monitor the focus by indicator and control place produces the reaction to the cameras to initiate recording. A successful method to accomplish this goal is to permit direct correspondence between these two contraptions, which can likewise build the range of potential applications.

The sight and sound content in cloud-based interactive media systems is fully accessible to clients for spilling or planning. Nevertheless, customers can not address the individual interactive media gadget or activate a range of procedures for organizing sight and sound gadgets, as these gadgets were not considered for cooperative correspondence. As a last point, the expense of sight and sound devices is still high,

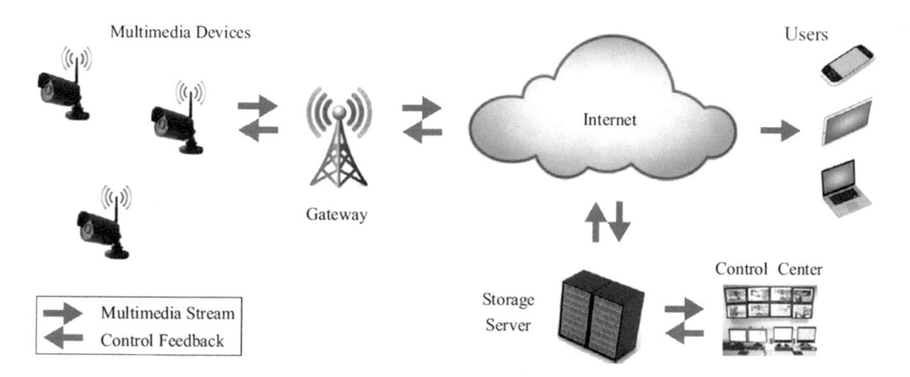

Fig. 1 Multimedia device architecture typical for wireless

limiting the sending of tremendous extension and huge use in everyday day-by-day life there. Presentations of the reference plan for Wireless Multimedia Sensor Networks (WMSNs), where blended media sources as a rule confine functionalities and transmit the substance through Wireless Sensor Network (WSN). The extent of WMSNs is deliberately confined to the sending circumstance in which the characteristics of the framework devices are known at the hour of course of action, right now gadget activity is obvious similarly as the QoS needs are pre-chosen at the individual contraption or at the framework level. Because of the fixed idea of WMSNs [11], the individual blended media devices are neither addressable nor outfitted with any setting-mindfulness or clear knowledge into the application, which is the reason a WMSN capacities as a single material inside the framework. Therefore, the mixed media gadgets in WMSNs need genuine heterogeneity in terms of their capabilities and capacities. In this way, the device process and QoS responsibility are not compatible with present system circumstances and preconditions for implementation.

The gadgets that communicate, transmit, and coordinate are required to be deeply specific in terms of their properties, communications capabilities, just as the mixed media material secures and prepares capabilities. Off chance we find the IoT systems that were designed and sent as late as possible. We see that few of the referenced highlights need to be updated in WMS and WMSN to completely integrate the sight and sound gadgets in the IoT environment. Figure 2 shows a reference engineering at the moment, often adopted by the European research company IoT-An [12]. The administration and asset layer is designed to provide functionality for disclosure and search, with the intention to things be not limited that providing information a solitary detailed vertical sending, but be open to exterior frameworks to support the relevant environment. Digital components are viewed with the goal of having a physical articles digital partner that enlarges items with additional cloud-actualized functionalities. Reviews, for example, include the representation of the functionalities of the objects just as previously created knowledge is put away and can be retrieved if necessary. In addition, item administrations can be connected to others from the

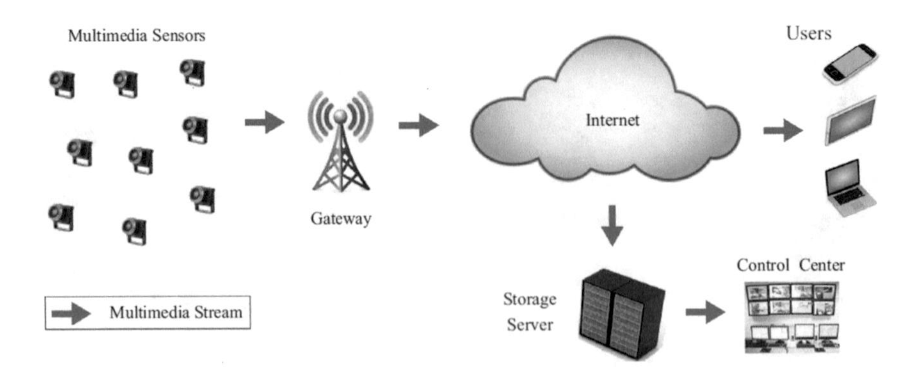

Fig. 2 Standard network architecture for portable, multimedia sensors

fourth layer of administration Web, so that specific applications can be focused on flying.

Table 1 shows relationship of unquestionable attributes of WMS, WMSN, and IoT as the eventual outcomes of this request. This likewise reports the features of the envisioned IoMT, which should be indistinguishable from those of IoT frameworks (aside from QoS and vital transmission limit); in any case, this does not imply that these features are legitimately open when introduced sight and clamor substance in IoT. Not at all like traditional media structures, IoMT-subordinate applications require severe QoS needs to give the great client experience. Many blended media frameworks might be touchy of incident; however, some others may require delay-sensitive correspondence. You can assemble the blended media applications into three gatherings, (I) spilling of sight and sound, for example, voice, video, and so on. (ii) Live spilling of blended media, and (iii) constant brilliant blended media correspondence. Regardless, the current TCP/IP Internet gives best effort service paying little mind to the type of material being conferred, and in this manner, QoS support is given (as much true to form) by application-level procedures.

In addition, to data transfer ability limit of IoT frameworks is frequently small, as effects are relied on either to supply scalar measurements of the observed condition or to obtain parcels of little directions. In an unexpected manner, the measurement of the information produced in IoMT can vary from not many kbps to a few mbps with the objective that the related frameworks should be flexible in the transfer speed offered. As of late, the accessibility of low-power sight and sound gadgets with minimal effort, such as Corresponding Metal Oxide Semiconductor

Table 1 Solution for comparison for existing for WMSN, WMS, IoMT, IoT

Parameters	WMS	WMSN	IoT	IoMT
Scope of applications	Limited	Deployment dependent	Dynamic and flexible	Dynamic and flexible
Node operation	Predefined	Predefined	Adaptive	Adaptive
Node resource and capabilities	Heterogeneous	Homogeneous	Limited heterogeneity	Heterogeneity
Energy efficiency	Not considered	Available	Available	Available
Scalability	Low	High	High	High
Interoperability	Low	Moderate	High	High
Deployment cost	Very high	Low cost	Low down cost	low down cost
Topology	Fixed	Limited	Ad hoc and dynamic	Ad hoc and dynamic
Bandwidth capacity	High	Moderate	Low	Low
Node operation	Predefined	Adaptive	Predefined	Adaptive
IP connectivity	Limited	Limited	Uniquely addressable	Uniquely
QoS multimedia	Available	Limited available	Unavailable	Available

(CMOS) cameras, CMOS-MEMS receivers, etc., has taken on tons of interest in remote sight and sound systems. As the principle goes, words usually cannot do justice to the image, sight and sound information provide extensive data that be able to adequately institutionalized in suitable organizations, sculpt and semantic representations depending on application setting to recognize and its valuable data.

Therefore, it is common for IoMT to have the ability of gigantic number of uses and it will be a simple piece of IoT. The sight and sound gadgets simply report the obtained mixed media data from their area in customary interactive media arranges. This sight and sound material could be distributed with the customer or put away in the cloud for handling and recovery on request [13]. For the most part, these sight and sound gadgets are intended to have the option to impart comparative qualities (homogenous gadgets), such as comparative correspondence stacks, comparable properties, and so on to other mixed media gadgets. In comparison, IoMT imagines media gadgets inspiring. For instance, cameras or amplifiers, to be universally accessible by a one-size-fits-all IP address with a similar soul to PCs and other Internet-related network administration gadgets. In addition, certain computing capabilities are built in the interactive media gadgets to render them enough interested in seeing the system and administration prerequisites and in triggering activities all by themselves. In this way, heterogeneous interactive media gadgets that obtain sight and sound substance from the physical condition, e.g., sound, video, images, etc., will impart and cooperate with each other just as with other keen things associated through the worldwide network cloud (Internet). This approach empowers the board with a broad range of uses the fields of house and construction computerization, plant observation processing, genius urban areas, transport, shrewd structure, and vitality [14].

Due to the complex decrease in the extent and expense of the sight and sound antenna gadgets, IoMT-based systems need to distributed everywhere where interactive media gadgets and other things are designed to frame impromptu communications with neighboring things as they travel around in the world system. Empowering of the self-sufficient association and self-administration of these extraordinarily unusual systems challenges by the solutions available that have been developed and created for unsurprising system designs. In fact, the mixed media knowledge from the different sensitive items reveals heterogeneous. Institutionalized representation should be provided to the functionalities and sensibly specific properties in order to consolidate coherently and to make the data procured into the context-conscious administrations for end clients. In any case, IoMT's recognition and acceptance or mostly IoT's resolve be sensibly incremental. While its perception and advancement necessitate significant up gradation in equally structures of the equipment just as the product arrangements do. We may describe the IoMT from the above discussion as the worldwide framework of interconnected media items that are curiously identifiable and addressable. Providing observed mixed media knowledge or operation like interfacing and talking to other sight and sound and not mixed media campaign and administrations, through or without straight individual mediation.

The combination of fundamentally heterogeneous sight and sound gadgets that form an exceptionally powerful and enormous device and also allow non-interactive media gadgets involves re-investigating the compositional plans and communication methods that have been used as of now. The procedure of IoMT-based aid able to be represented with assist of a four-arrange IoMT design, whereby every particular phase comprises a consistent arrangement of the procedures performed at each stage of a comprehensive support. The projected IoMT structural design reflects different points of view, imperatives, and difficulties at each of these phases to recognize sight and sound communication in IoMT. In the following areas of this paper, the functionalities of each of these individual phases are widely discussed. Right off the bat, the mechanisms for securing and encoding sight and sound information are covered. The behavior of media detecting gadgets and their attributes is especially investigated.

3 IoMT in Multimedia Detecting

The IoMT is increased the IoT, where single of the primary destinations to facilitate video gushing as major component IoT recognition. In the IoMT, asset-related heterogeneous low down power mixed media gadgets able to communicate with every supplementary and with different IP addresses available. A similar soul as related gadgets for the PCs and other systems administration through the Internet. The difficulties posed by IoMT are, for example, like IoT, handling a lot of data, questions, and calculation as well as certain unmistakable necessities. The interactive media gadgets should be small measured objects fitted with a limited measure of strength assets in IoMT-based remote sight and sound systems, which they need to use effectively to create arrange lifetime. Good approaches for vitality should therefore have been established to coordinate regulatory methodology. Essentially, mixed media gadgets must be incorporated with relevance and careful information setting, consequently that mixed media satisfied from the objective condition can be collected if required, thus restricting the excess data obtained.

For the most part, battery-fuelled portable interactive media devices are relied upon. Since, the securing of sight and sound and its handling are systems which expend very much on energy. It is thus considered that interactive media sensor systems should be designed to extract as much vitality as possible from the earth. Therefore, self-fuelling mixed media sensor gadgets are figured in IoMT worldview to harvest vitality from various sources of vitality in the device field. Accordingly, in [15] organizing numerous sources of vitality, such as sun-powered cells, piezoelectric, thermoelectric and radio-energizing devices, vibration, warm and radio-frequency vitality selection is advanced. The possible gathered vitality from these sources is shown in Table 2 alongside their individual efficiencies. Regarding batteries, the sight and sound devices should be fitted with different sources of vitality, such as sun-based cells, thermoelectric collectors, etc. In cruel circumstances where battery replacement is not down to earth, this will outcome in a considerable increase in the ease of use of IoMT. One of the important aspects of media information gathered is the encoding of the incomplete sight and sound data detected.

Table 2 Typical technique for extracting energy

Source	Source of energy	Source power	Harvested power (uW/cm^2)	Efficiency (%)	Issues
RF-CSM 900 MHz	RF harvester in urban	0.3–0.03 uW/cm^2	0.1	~50	Coupling and rectification
RF-GSM 1800 MHz	Environment	0.1–0.01 uW/cm^2			
Thermal	Human body heat dissipation in air	20 mW/cm^2	25	~0.1 to 3	Small thermal gradients; efficient heat $inking
Vibration	Human walking with harvester on the floor	0.5 + 1 m/$1 + 50 Hz	4	~25 to 50	Variability of vibration frequency
Photovoltaic-indoor	Harvester inside a building environment	0.1 mW/cm^2	10	~10 to 24	Conform to small surface area
Photovoltaic-outdoor	Harvester open air in a sunny day at soon	100 mW/cm^2			Wide input voltage range

Postpone intended for sight and sound spilling: Multipath blurring debases different neighboring bits appropriate to repeated deep blurs. Interleaving was done at the transmitter so that errors are distributed separately when de-interleaving the receiver. Be that as it may, letting separations at the transmitter overlap and restoring calculations at constant postpone bound, as far as possible. The methods for video coding are classified into three broad classes, I standard video coding, (ii) video coding transmitted, and (iii) compressive detection when considering these necessities.

4 Major Security Challenges Related with IoT Layered Design

By using advanced mobile phones and portable tools, IoT makes it easier for individuals to screen, monitor, inspect, and manage various exercises from anywhere. Ruptures of the respectability, legitimacy, and privacy of the collected data will certainly occur. What is more, the information captured could be misused for different purposes. In the large communications, it is commonly reported that mechanical vehicles like transportation, buses and ambulances, power lattices, and even child screens have been off-traded. Therefore, there is a need for ensuring the data security and the gadgets themselves on every single layer of the IoT.

Use several similar devices, heterogeneous conventions, and stages across the board produces numerous new security challenges in terms of protection, classification, confirmation, control, and confidence [16]. These gadgets and applications contain data, some of which can potentially undermine association and people's security. Portions of the perceptible security challenges are simplified in Fig. 3, arising from the heterogeneous state of IoT. IoT architecture has three basic layers to follow:

1. The layer of awareness also called the availability of sensors and the network layer;
2. The layer of transportation also called the interface and device layer; and
3. The layer of operation at the top of the stack. Security risks and ruptures are simplified at each layer and shown in the following section.

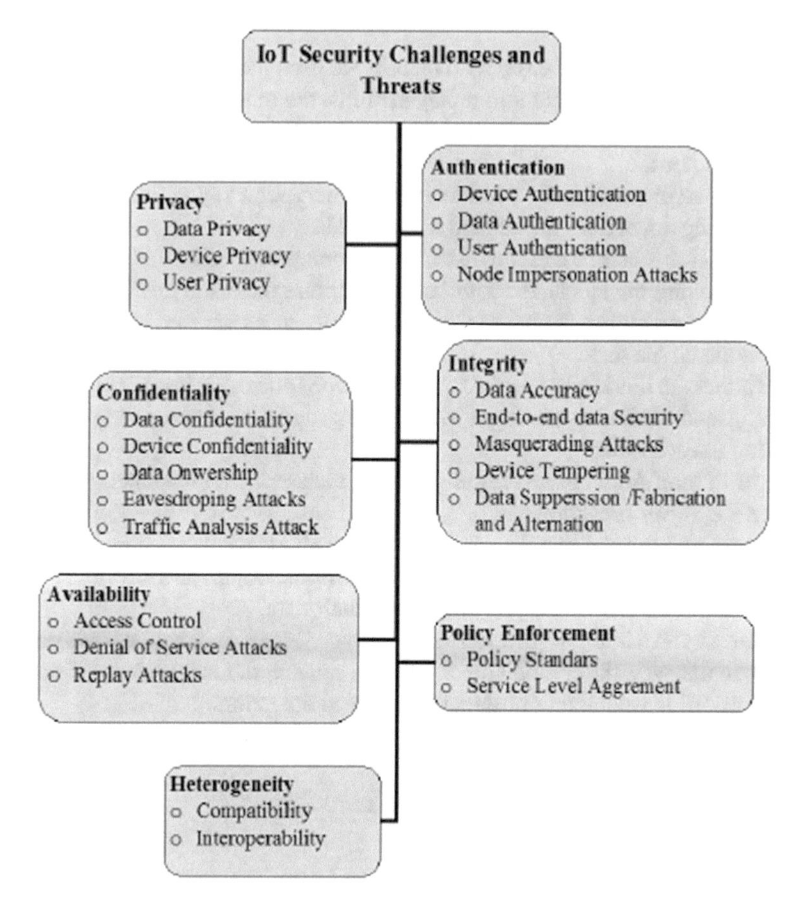

Fig. 3 Big protection and IoT environmental threats

4.1 Security Dangers on the Perceptual Layer

The Perceptual layer is the base layer of the IoT stack which includes sensors and intelligent apps [17]. This layer provides accessibility to the sensor and administration of the systems. This is where continuous data is collected and treated in Fig. 3. Each layer has two sub-layers to it. The main sub-layer is sight and sound that includes video and sound details. The corresponding sub-layer is image and object containing information and photos about the material. For example, body sensors, ecological sensors, and reconnaissance sensors are assembled by their motivation and information forms. First come the assaults that can smash the authentication, classification, and integrity of knowledge and gadgets on the layer of discernment.

- Invasive attack and hardware tempering: Since most gadgets are sent out and vulnerable, an aggressor may pursue and access them without much of a stretch and alter and temper the gadgets or remove the sensor to collect the data and rebuild them afterward.
- Replay attack: In an IoT case, RFID labels are used mainly as regards computerized data trading. Therefore, in replay assaults, the transmission signals between the RFID labels and the peruse are interrupted and replayed when the peruse asks for any request.
- Timing attack: An attacker will identify the encryption key in this kind of attack by breaking down the time needed to encrypt it.
- Node capture attack: The ID data of a tag is compromised in this kind of attack. By controlling the labels, the assailant can capture the distinguishing proof. The aggressor then clones the tag to sidestep security measures and is able to control and clone all the data.
- DoS attack: In this kind of denial of service (DoS) attack, the assailant can continuously send vindictive messages to stick the gadgets, thus preventing them from talking to each other.
- Server failure: Another vulnerability that can cause incredible damage in an IoT system is server breakdown.
- Malware and bot net attacks: Assaults based on malware. In these attacks, the assailant can create malicious codes such as worms, trojans and infections to taint and monitor the clever gadgets to separate their data.
- Complexity: An IoT system consists of various devices, innovations, and conventions. In this way, the complexity of the large number of devices, innovations, and conventions is yet another obstacle to its broad reception.

4.2 Security Dangers on the Transport Layer

The 'unit layer' is also called the transmission layer. This layer is responsible for the transmission to the applications or end clients of data collected from sensor gadgets or clever objects. For the most part, the data is transmitted via remote channels of communication using innovation such as Bluetooth, Wi-Fi, Zig Bee, and 4G cell remote. Due to various encryption techniques, key administration calculations, and character validation, remote correspondence is currently increasingly safe. Nonetheless, aggressor will violate this layer in the accompanying ways in any case:

- DoS attacks: A DoS attack on the vehicle layer can cause significant rates of data traffic and malicious blockage of channels by caricaturing, hi-flood assaults, and resynchronization assaults to disrupt network gadget assets.
- Eavesdropping: By listening stealthily, the noxious consumer reaches the network and catches the data flowing through the device, thereby fully breaching the individual data privacy and confidentiality.
- Passive monitoring: Another immense security hazard is passive observation of the IoT arrangement. A programmer can screen and decode an IoT gadget's ongoing traffic over the system and change the ID rules and banner suspect activities. The transport layer has a profound powerlessness against latent assaults. Man in center assault is a case of inactive assaults in which an intruder can monitor the channel of communication and thus imitate the fake data.
- Issues of compatibility: The objects impart in the IoT setting in the state of elements where there are specific conventions for different things at different times and circumstances. An attacker will manipulate the system's dynamic nature and collect the data for use in criminal matters. Similarity of devices and conventions is thus one of the key safety hazards to mitigate.
- Identity theft: Another security problem is fraud, where unapproved or malevolent clients can sniff the authentication data, such as a hidden word and EPC code and gadget keying techniques, to sniff the validation data. This can lead to sensitive data and imperil people's security.
- Routing attack: Another security risk is a steering attack where IoT data can be seized or mocked.

4.3 Security Dangers on the Application Layer

The application layer is the main IoT interface layer. It offers the requisite types of assistance to the end client and decipheres the data gathered at the observation layer from the shrewd object. In the application layer, different areas of application are planned, for example, human services, research, transport, retail, and much more. The customer can access and monitor these applications anywhere, any time, and from any point. The health risks associated with the application layer shall be as follows:

- Runtime attacks: In runtime attacks, an attacker gathers or changes basic data through unauthorized access to and violation of the protection and accessibility of properties that can cause crushing knowledge misfortune and can even endanger human lives.

- Malwares: When updating the use to include new functionality or highlights, an aggressor may be able to abuse the application by infusing malignant programming, such as worms and trojans.
- Privacy threat: The IoT framework transmits various personal and private details, such as the area of a document, people's actions and their social relationships, or an association's registered information that may pose a privacy danger to their clients. An attacker may extract this data for offences, or the data may be made available for advertising purposes to invested persons.
- Integration: There are tons of security threat emerging as IoT is becoming more and more well known. For example, brilliant house, knowing testing, system observation, shrewd transportation, and observation, the distinctive kind of uses and administrations have various security needs. In this way, the coordination of different procedures and safety criteria is needed into persons IoT applications.
- High overhead: Many IoT devices have low computational power, but huge amounts of information collected from the earth around render overhead that can obstruct the correspondence and undermine the usability of the necessary applications. A middleware is now required to store, process, and start finishing transportation in order to reach the administrations and application. Middleware's information security, confidentiality, and reliability are seen as another fundamental concern.
- Phishing: In a phishing attack, for example, the attacker sends messages from the undermined phones, cameras, home appliances, and shrewd vehicles, imitating a real gadget and arranging what looks like actual messages to the query. The attacker can change the recognizable proof and validation certifications and the misuse of the data for crimes.
- DoS attacks: The code layer conveys programming for different applications, thus being helpless against an asset exhaustion attack. Aggressors can constantly send demands by overwhelming the network with countless parcels to disrupt the functionality of the application.
- Social engineering: In social building, aggressors exploit social relationships and obtain data efficiently, and harm an individual or association's security. In the section below, assaults that may compromise sight and sound communication are discussed when all is said and done. After increasingly widespread assaults, a portion of these assaults actually represents a particular danger to sight and sound correspondence.

5 Security of Multimedia in IoT

For example, Web-based gaming, video conferencing, video recognition, and much more, there are numerous digital media bringing together applications and administrations running around the world system in Fig. 4. For example, the IP-based gadgets, IP cameras and wearables make the computer a match for the gadget. A

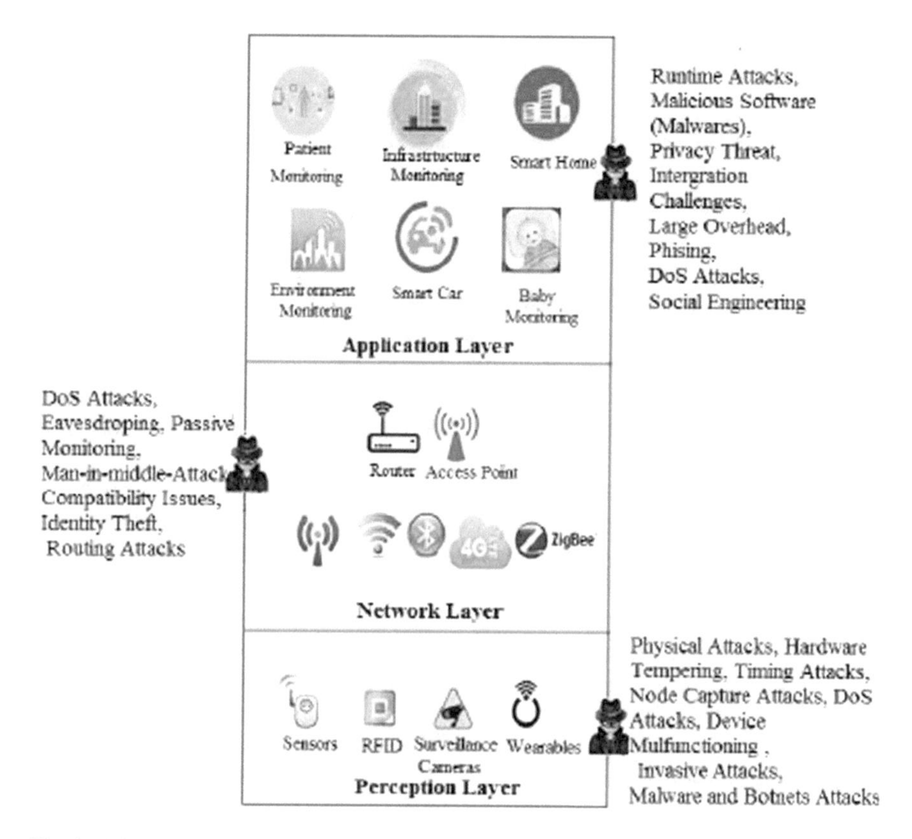

Fig. 4 Vulnerabilities and attacks in every IoT architecture layer

reality and now remote surveillance of children, continuous tracking and observation, and monitoring of objects and individuals is possible. Due to the high use of sight and sound applications throughout the world IP traffic, enormous movements of mixed media traffic have occurred since late. As indicated by CISCO trends and estimate study, in the coming years, multimedia traffic will completely control IP traffic on the network. Sight and sound is a sub-layer of the layer of discernment within IoT layered engineering. The message of sight and sound or video stream, captured from gadgets (for example, cameras) is sent. Storage servers for storage and planning due to their limited memory limit and from that point through the Internet to the end client. In doing so, the video goes through a few hubs/passages to access such repositories using heterogeneous communication advances. Nonetheless, these pathways and middle of the road structures can be surprisingly defenseless to attacks. Some of the media related threats are listed below:

- Cloning: The assailant imitates and substitutes the multimedia in a cloning assault and infuses vindictive data, such as falsified images, so that the real consumer may receive fashion data.

- Area disclosure attack: In this assault, the malevolent customer gathers information about the locale and air or coordinates the thing's reference information from the caught video and may utilize this information for aggressor mental activities.
- Node impersonation attack: The culprit goes about as the genuine individual right now and sends mock messages to different center points setting off the breakdown of various center points. Attacker can likewise reconfigure the system and expel safety efforts to complete extra ambushes on center points so recipients can get to basic information.
- Eavesdropping attack: The attacker takes video streams during transmission over an understanding and gets helpful information and disregards the information security.
- Resource consumption assault: By transmitting long bundles to the target device, an aggressor can manipulate these confines due to power and usefulness imperatives.
- Traffic analysis attack: In the attack, an aggressor analyzes behavioral instances and characteristics of the video stream recorded as a tool for all those who are more likely to be affected, due to the quantification of the policy.
- Unlawful recognition: In this safety hazard, hoodlums may have the option of blocking information and screen traffic from observation gadgets and mid-range gadgets to identify people under recognition and the incentive for recognition.
- Disguising attack: The assailant utilizes the character of the genuine center right now creates distorted message in the system.
- Replay attack: In a replay assault, the video could be pre-recorded by the culprit who put this recorded video in a perception association at that level, so the control room managers would think about this video as being ceaselessly sent.

Accordingly, it appears to be seen that there are various possible attacks on interactive media capturing gadgets and messages, for which the following section hardly proposes any cures.

6 Proposed Interactive Media Security Engineering in IoT

The interactive media devices collect a great deal of information from the environmental factors they are sent to. The information gathered (for example, recordings) demands huge memory assets and is hungry for transfer speed. In this way, low-power IoT asset compelled gadgets require memory assets and transmission ability to process and transmit the data. It is therefore necessary to ensure the security of sight and sound and the confidentiality, trustworthiness, and privacy of data by the customer in a viable manner. As of now, various protection measures are used, such as stenography, coding, watermarking, and interactive media pressure. Must ensure that these all devices are not sufficient in IoT condition due to various limitations on gadgets over the mixed media yet. In general, Dazzle/Naïve encryption

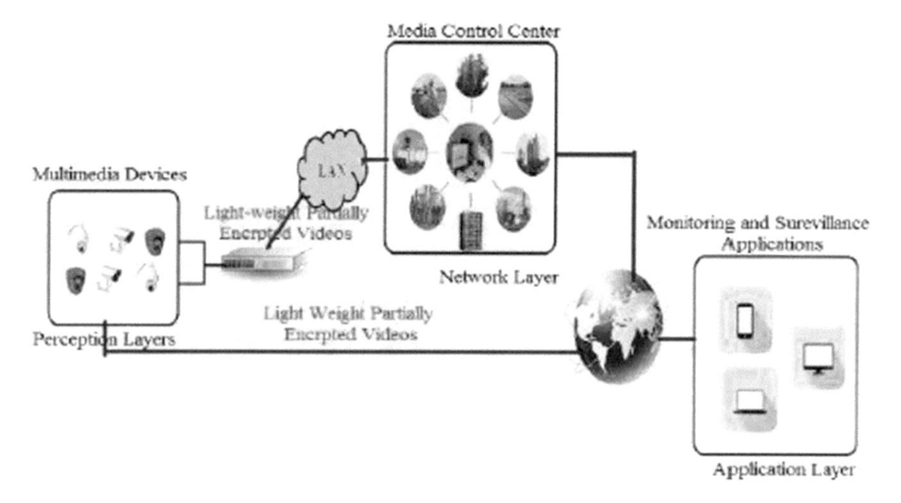

Fig. 5 Secure multimedia architecture in IoT environment

is used to ensure the transmitted data. Daze encryption means the complete encryption of mixed media information such as header data with payload. This encryption causes numerous problems, such as massive computing costs, encoding complete information, requiring more transfer speed for the transmission of scrambled information, and the most important thing is that sight and sound information can not be translated/transcoded in conjunction with the related IoT gadget data (Fig. 5).

Recalling each of these problems, the paper proposes several lightweight halfway encryption techniques to refresh sight and sound protection at different levels. The whole video will not be scrambled in proposed phased fractional encryption methods, just not many hidden bits of the video are proposed to decrypt. Those proposed levels of security provide the remedy for mixed assaults on media security. The proposed sight and sound protection concept in IoT condition is being extended in IoT engineering as of now.

Most of the information collected by the mixed media gadgets and sensors will be scrambled and then transmitted over the network in the proposed security multimedia engineering, so that the end clients will receive the encoded recording. Henceforth, if any intruder triumphs to block the video, the video can be seen as mixed in structure. Another advantage of the new encryption proposals is that they will preserve the encoding/deciphering highlights of videos, so the decoders would not crash after the plans have been implemented. As the word lightweight encryption indicates, video details will not be scrambled to the full. Clearly transparent components of the caught/transmitted/released video are encoded in such a way that the video can be identified in a mixed structure as opposed to being watchable. Different levels of security encryption are provided to secure mixed media information from assaults and ensure that individuals and spots are secured during observation and inspection.

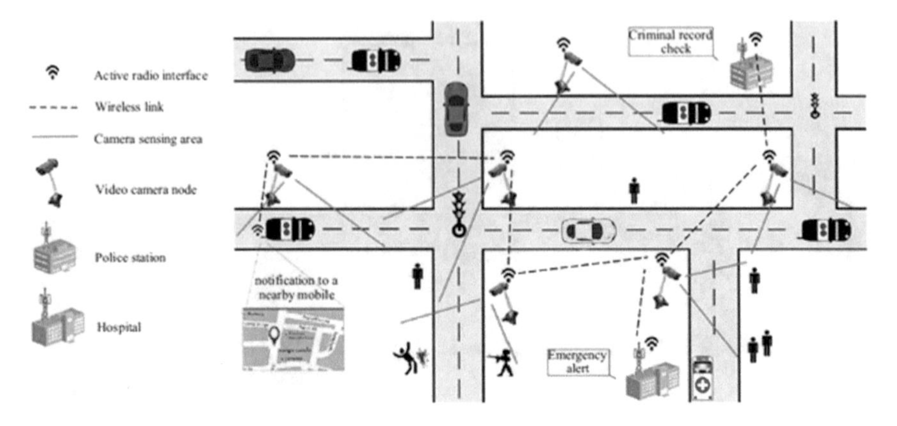

Fig. 6 Automated public security

7 Automated Open Security

In the IoMT worldwide video streams, transparent health can be encouraged and one of the conceivable situations is presented. A person has fired a shot on a packed lane. Sound analysis by the cameras in the shooter area will promote the detection of shots. The shrewd mixed media gadgets sent in the area will identify the incident of an unusual occasion to send the individual in question swift clinical support. The incident data is accounted for by an emergency clinic which immediately dispatches the rescue vehicle to a close. The specialists are encouraged to photograph the continuous spill of video from the episode, thereby civilizing the scientific reply time. The gunman would be recognized between the party all the while, and his motion would be expected. With the shooter's movement bearing in mind, camera organizer must update the following cameras clicking the way of change the objectives and course (Fig. 6).

The controller must speak to the adjacent watch for regulation enforcement and incessantly monitor the shooter area along the watch's path. In the meantime, the organizer will forward the data to personality databases, regulate, and clinical databases with the intention of verifying his identifying facts, past criminal records, and clinical history. The route on the watch must speak to the traffic signs to track traffic progress so as to stay away from any breaks in its growth.

8 Conclusion and Future Work

Via brilliant gadgets and interactive media apps, there is an evolving trend toward IoT situations that offer remote day-to-day checks. Monitoring devices have been included in various open and private groups, organizations, and spots that range from remote monitoring of areas affected to assessing the status of young people at home.

Nevertheless, the use of interactive media technologies for remote observation across the board is also growing as a genuine risk to associations and people's stability, prosperity, and well-being. The classification and protection of the information collected from the sight and sound gadget is important in media applications.

References

1. Ekram, H., & Bhargava, V. K. (2007). *Cognitive wireless communications networks*. Springer Science & Business Media.
2. Ganesh Babu, R., & Amudha, V. (2016). Cluster technique based channel sensing in cognitive radio networks. *International Journal of Control Theory and Applications, 9*(5), 207–213.
3. Ganesh Babu, R., & Amudha, V. (2016). Spectrum sensing cluster techniques in cognitive radio networks. *Procedia Computer Science, 87*, 258–263. (In *Proceedings of 4th International Conference on Recent Trends in Computer Science & Engineering, (ICRTCSE) in association with Elsevier*).
4. Ganesh Babu, R., & Amudha, V. (2018a). Allow an useful interference of authenticated secondary user in cognitive radio networks. *International Journal of Pure and Applied Mathematics, 119*(16), 3341–3354.
5. Ganesh Babu, R., & Amudha, V. (2018b). Comparative analysis of distributive firefly optimized spectrum sensing clustering techniques in cognitive radio networks. *Journal of Advanced Research in Dynamical and Control Systems, 10*(9), 1364–1373.
6. Ganesh Babu, R., & Amudha, V. (2018). A survey on artificial intelligence techniques in cognitive radio networks. In *Proceedings of 1st International Conference on Emerging Technologies in Data Mining and Information Security, (IEMIS) in association with Springer Advances in Intelligent Systems and Computing Series* (pp. 99–110).
7. Ganesh Babu, R., Karthika, P., & Aravinda Rajan, V. (2020). Secure IoT systems using Raspberry Pi machine learning artificial intelligence. In *Proceedings of Second International Conference on Computer Networks and Inventive Communication Technologies. Lecture Notes on Data Engineering and Communications Technologies* (Vol. 44, pp. 797–805). Singapore: Springer.
8. Karthika, P., & Vidhya Saraswathi, P. (2019). Image security performance analysis for SVM and ANN classification techniques. *International Journal of Recent Technology and Engineering, 8*(4S2), 436–442.
9. Karthika, P., & Vidhya Saraswathi, P. (2020). Raspberry Pi: A tool for strategic machine learning security allocation in IoT. In N. Raju, M. Rajalakshmi, D. Goyal, S. Balamurugan, A. Elngar, B. Kesawn (Eds.), *Empowering artificial intelligence through machine learning* (pp. 133–141). Apple Academic Press, CRC Press, Taylor & Francis Group.
10. Karthika, P., Ganesh Babu, R., & Nedumaran, A. (2019). Machine learning security allocation in IoT. In *IEEE International Conference on Intelligent Computing and Control Systems*, Madurai, India, 474–478.
11. Mitola, J. (2000). *Software radio architecture: Object-oriented approaches to wireless system engineering*. New York: Wiley.
12. Nedumaran, A., Ganesh Babu, R., Kassa, M. M., & Karthika, P. (2019). Machine level classification using support vector machine. In *AIP Conference Proceedings of International Conference on Sustainable Manufacturing, Materials and Technologies (ICSMMT 2019)* (Vol. 2207, Issue 1, pp. 020013-1–020013-10), Coimbatore, India.
13. Rondeau, T. W., & Bostain, C. W. (2009). *Artificial intelligence in wireless communication*. USA: Artech House.
14. Maurya, S., & Mukherjee, K. (2019). An energy-efficient architecture of IoT based on service oriented architecture (SOA). *Informatica: An International Journal of Computing and Informatics, 43*(01), 87–93.

15. Maurya, S., & Mukherjee, K. (2017). An energy efficient design of cloud of things (CoT). *Journal of Information and Optimization Sciences (JIOS), 39*(1), 319–326.
16. Maurya, S., & Mukherjee, K. (2016). A novel method for secured cloud-health-care-system. *International Journal of Control Theory and Applications, 09*(19), 9291–9303.
17. Maurya, S., & Mukherjee, K. (2015). A literature survey on mobile cloud computing. *International Journal of Applied Engineering Research, 10*(79), 329–334.

A Robust Approach with Text Analytics for Bengali Digit Recognition Using Machine Learning

Dipam Paul, Prasant Kumar Pattnaik, and Proshikshya Mukherjee

Abstract Vernacular digit or text recognition has always been a challenge courtesy of the undecipherable images and uniqueness in every human being's handwriting. In this paper, we have aimed to classify the Bengali numerals obtained from the NumtaDB dataset with the implementation of a convolutional neural network (CNN). Moreover, we have also conducted a survey based on Bengali numerals and performed a sentiment analysis on the recorded responses of our subjects using the same classification technique and thereafter have touched upon an area of application of vernacular digit recognition which has seldom gone unnoticed. We have also endeavoured to provide an in-depth literature survey of the works that have been done in this area previously. Our classification model gives us a whopping accuracy of 98.40%.

Keywords Text analytics · Bangla digit recognition · Augmentation · Classification · Sentiment analysis

1 Introduction

India is a country where multiple languages are spoken, written and are used to converse with. There are 22 languages which were recognized by the Constitution of India in its eighth schedule; these are: Assamese, Bengali, Gujarati, Hindi, Kannada, Kashmiri, Konkani, Malayalam, Manipuri, Marathi, Nepali, Oriya, Punjabi, Sanskrit, Sindhi, Tamil, Telugu, Urdu, Bodo, Santhali, Maithili and Dogri. Among these, the two most spoken languages in the India subcontinent are Bengali and Telugu.

D. Paul · P. Mukherjee (✉)
School of Electronics Engineering, KIIT DU, Bhubaneswar, India
e-mail: pmdonamukherjee7@gmail.com

D. Paul
e-mail: dipampaul17@gmail.com

P. K. Pattnaik
School of Computer Engineering, KIIT DU, Bhubaneswar, India

R. Kumar et al. (eds.), *Multimedia Technologies in the Internet of Things Environment*, Studies in Big Data 79, https://doi.org/10.1007/978-981-15-7965-3_10

Furthermore, there exist twelve Indian scripts, namely English, Bangla, Devanagari, Gujarati, Gurmukhi, Kannada, Malayalam, Oriya, Tamil, Telugu, Kashmiri and Urdu [1]. The script which is used for Hindi, Sanskrit, Marathi, Rajasthani and Nepali languages is the ancient Devanagari script, and the one which is used for Bengali, Assamese and Manipuri languages is called the Bangla script. Pattern recognition as a subject encompasses a wide array of topics ranging from text recognition, gesture recognition finding faults in a machine or medical diagnosis, etc. Handwriting recognition is a concentrated subset under the sphere of pattern recognition where we utilize and apply statistics and other information processing techniques to draw inferences from images gathered from raw data and further process them and group them into different classes each containing a separate set of elements. This is essentially one of the main applications of pattern recognition which is to extract common features from a given set of classes and then categorizing them and hence recognizing the pattern.

Bengali Digit Recognition is a remarkable starting point in developing an OCR. With the advancement of machine learning and deep learning, the avenues and possibilities for solving problems using and involving computer vision have increased significantly. These new methods have been proven to be very useful especially in the domain of Text Analytics, which is to decipher numerals and alphabets from a given set of handwritten digits and alphabets. The use of Text Analytics is very instrumental in the present paradigm as one of the more primary applications of it helps us to determine and decipher texts from ancient scriptures while it also decreases the human effort to learn new languages in order to operate at an organizational level, thus negating and deleting the language barrier problem which is prevalent.

The main objective of this paper is to recognize and classify unique Bengali digits wherein, there is an inherent challenge which is the fact that every human being has a different style of writing. Our contribution lies in tackling the aforementioned task and thereafter approaching the challenges and bringing out a model with robust performance and high accuracy for a huge, unbiased and highly augmented NumtaDB dataset. We have used various image processing techniques, and then we have classified the numerals based on a deep neural network architecture model.

1.1 Need for Text Analytics

It is needless to say that in the recent years we have all witnessed the role of Text Analytics in the world, especially with new and emerging technologies such as optical character recognition (OCR), optical mark recognition (OMR) and handwritten character recognition (HCR). This new methodology becomes very significant because part of the problem is a huge amount of unstructured text especially in the corporate sector, and Text Analytics helps you to uncover and unravel all the hidden patterns and themes so that your client knows exactly what are the demands, expectations or even complaints of his/her customer. Therefore, it is indeed a challenge to find the connotations in a given set of data which are unstructured and using these Text

Analytics tools gives us a much better understanding of the data we are dealing with which is not easily quantifiable in any other way. Bengali, being the second most spoken language of India, fifth most spoken in Asia and is also in the top ten most spoken languages of the world. Therefore, this denotes the importance of digitizing of Bengali numerals for smooth flow of communication while highlighting the motivation of this paper.

1.2 Challenges

Barrier of language: As mentioned above, the primary challenge in this research is to tackle the problem of unstructured data and data which are essentially text-heavy with the utilization of Text Analytics tools. On the other hand, this system also addresses one of the more age-old problems since the time of civilization, which is that of language being a barrier in communication. Moreover, language, in general, is ambiguous and therefore, this could be also classified as a step towards eliminating that ambiguity and help us assess a problem in a better way and facilitate a solution-oriented approach for the same.

Dataset acquisition: In this research, we found a stark shortage of a clean image dataset containing unbiased, completely randomized numerals. Distorted images often mislead us to false results; hence, one needs to be careful while choosing a dataset for their model to train upon.

Unique handwriting of individuals: The most pressing challenge here was the inherent variability of handwriting which keeps changing with every subject we take. Therefore, any discrepancy in the same could also mislead us to inconsistent results, which would rather prove to be inefficient.

Broken character: This is often seen in most datasets wherein the characters in our image set tend to be broken or improper which could be possible due to any reason, for example, the broken tip of pen or pencil.

Touching digits: In this research, while addressing one of the hurdles, as mentioned above, it was to find clarity in the images given. We often see that two digits touch each other and that could lead us to falsified error which would again be detrimental towards the entire study. The problem largely occurs where we segment the images and therefore should be avoided as addressed by Bathla et al. [2].

Overlapping aspect: This kind of inconsistency occurs only when there are unwanted marks or strokes in the paper in which the dataset has been collated, and the neural network often detects this malicious inputs and tends to digress away from the result we desire.

Skewness, font size and style of writing: Alignment of the digit towards the left or right side of the projection again results into erratic inference. This can lead to overlapping as well, hence is at odds with the entire structure of our study and thus, was avoided as investigated by Kumar et al. [3].

1.3 Applications

In rural areas, especially in the eastern part of India, Bengali is one of the most spoken and used language even at an official level, such as communication via post offices, and these processes could be automated using this algorithm incorporated with a few minor tweaks as per the community's demands. Similarly, identification is another big issue when it comes to these rural areas, and therefore, digit recognition invariably would play a very instrumental role in identifying these parts of communities where this problem is persistent for example, half-torn, undecipherable ration cards are a common observation and could be mitigated using the approach we follow with Text Analytics.

2 Literature Review

In 2005, S. Basu et al. proposed the Dempster–Shafer (DS) technique for handwritten Bangla digit recognition purpose. Here, the authors have used DS with MLP classifier for classification purposes. Here, the test accuracy rate is 95.1% [4].

In 2006, U. Pal et al. proposed another scheme for offline Bengali handwritten numeral recognition purpose. Here, they have used water overflow from reservoir concept. The test recognition accuracy rate is 92.8% [5].

In 2009, U. Bhattacharya et al. presented a classifier model and they used handwritten numeral Devanagari database. Here, they have used a multistage cascade recognition scheme. Here, the test recognition accuracy rate is 98.20% [6].

In 2012, Y. Wen et al. proposed a Bengali handwritten numerical recognition technique by using the classifier model. This method solved the large dataset dimension problem. Here, they have used Bayesian discriminant with kernel approach with UCI and MNIST dataset. The recognition accuracy rate is 99.08% [7].

In 2013, A. L. Mass et al. proposed that activation of ReLU in all layer. Here, they considered randomly moving dataset and each image is rotating between 0 and 50 degree and vertically shifted by 0–6 pixels. SCM, SCMA and ACMA are the three types of setup here they considered, and the test accuracy rate is 99.50% [8].

In 2015, M. M. Rahaman et al. proposed CNN-based model. Here, author normalized written image characteristics. The test accuracy is 85.36% [9].

In 2016, M. A. H. Akand et al. used simple rotation model. Here, they considered ISI handwritten model. The test accuracy rate is 98.45% [10].

In 2017, Md. Shopon et al. used $32 \times 3 \times 3$ kernel layer division. Here, an autoencoder was used for unsupervised pretraining through deep CNN [11].

In 2018, A. Choudhury et al. proposed colour histogram and oriented gradient histogram for numerical item selection SVM that are used for the input form. CMAERDM is used for data selection. Here, the test accuracy is 98.05% [12].

In 2019, Abu Sufian et al. proposed BDNet wherein a densely connected CNN model was used to perform on the ISI Bengali handwritten numeral dataset. Here, the test accuracy was 99.775% [13]. Text Analytics for NumtaDB dataset (Fig. 1).

2.1 Dataset

The dataset we used named NumtaDB consists of six different datasets procured from different instances of events. The datasets are unbiased with regard to geography, demography, gender and age. It consists of more than 85,000 images, and these were all checked and evaluated religiously such that all the digits could be made at least readable to one person hence the data acquisition sets a benchmark when compared to other Bengali numeric datasets. Our train data matrix size is: [(48, 48), (72,045, 48, 48), (72,045, 10)], and our test data matrix size is: (17,626, 48, 48) (Figs. 2 and 3).

2.2 Image Resizing and Alignment Tuning

In the dataset, the training and testing sets will have different subsets of data ranging from ‘a’ to ‘f’—this will depend on the root path. Here, we set the parameters as RESIZE DIM = 48 and CHANNELS = 1, and hence, all the images will be resized to 28×28 pixels. We do this to avoid any potential problem of errors due to alignment or positioning of text when treated by a neural network. Now, for the alignment tuning part, we set FIG_WIDTH = 28—which corresponds to width of each figure and HEIGHT_PER_ROW = 3—which entails height of each row when showing a figure which consists of multiple rows and thereby we also set the vertical and horizontal alignment to ‘centre’ for better feature extraction and proper classification to take place (Fig. 4).

2.3 Image Augmentation

While we acknowledge the praise and popularity that the deep learning algorithms have earned over the years with regard to problem-solving especially in the periphery of image processing, we also take into account a very fundamental and persistent problem that has always been present alongside these algorithms, which is the requirement of a huge amount of data for the model to work. Therefore, we use image augmentation techniques in our study and try to replicate images from the given set

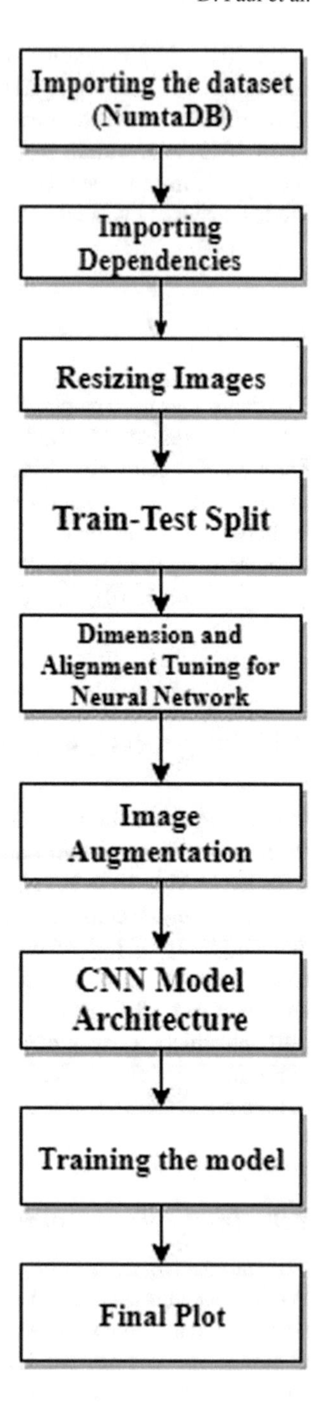

Fig. 1 Proposed method

Fig. 2 Data collation sample #1

of images and then incorporating the same in our model. In this process, we attempt to obliterate the requirement of an extensive data acquisition process which is practically impossible at times. We also incorporate various image processing techniques in this step and add blur, noise and filters such as—Gaussian blur, additive Gaussian noise, contrast normalization, affine and piecewise affine among others. In doing so, we augment the images after processing them with the aforementioned instruments and thereby we process 160,045 images here. Consequently, we observed that this part of the code had the longest run time. We also observed that after augmentation, few of the images have distorted pixels, which was inevitable; we will now eliminate those unwanted images from our CNN architecture, which is elucidated in the next subsection (Fig. 5).

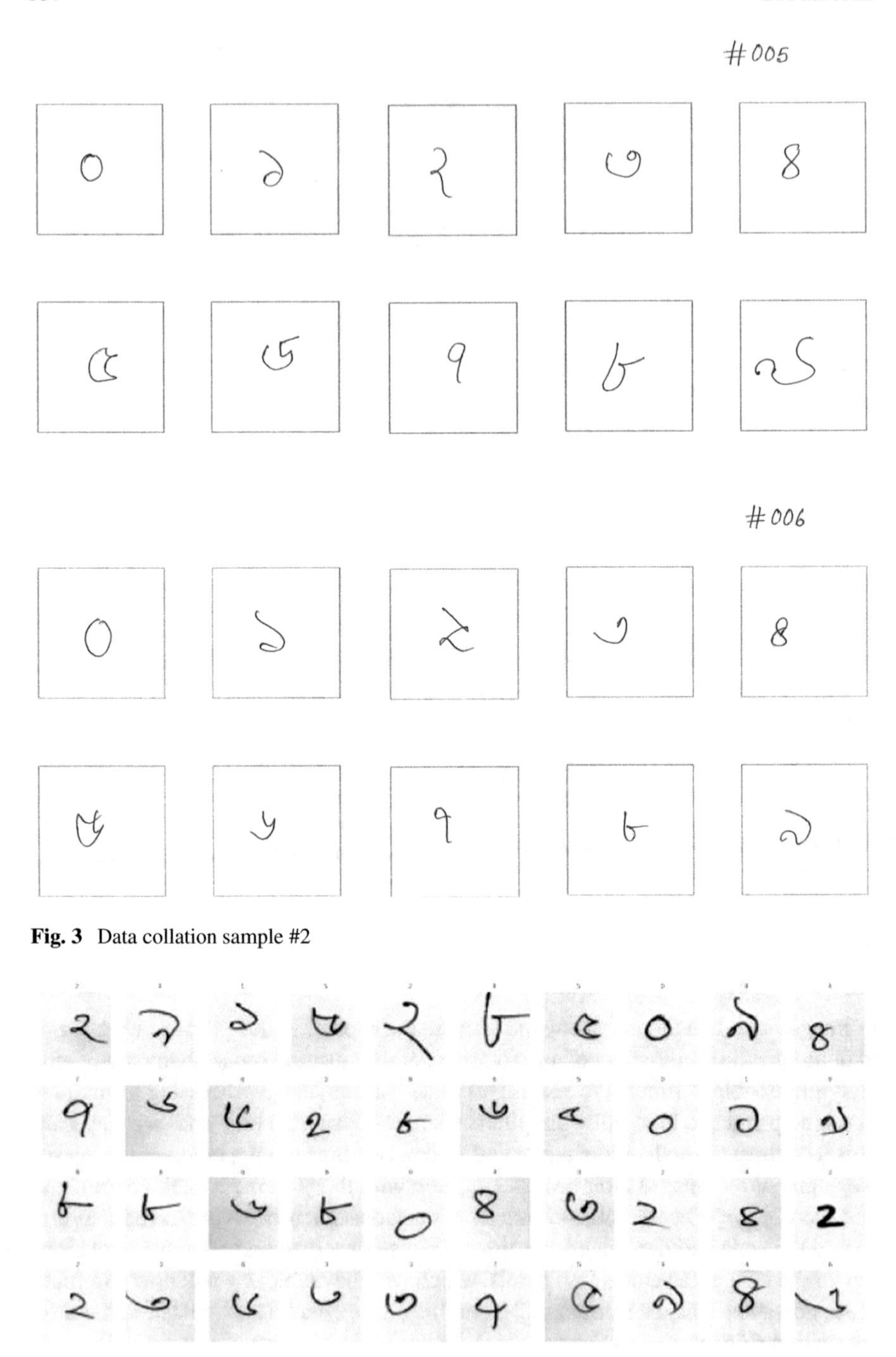

Fig. 3 Data collation sample #2

Fig. 4 Resized training images

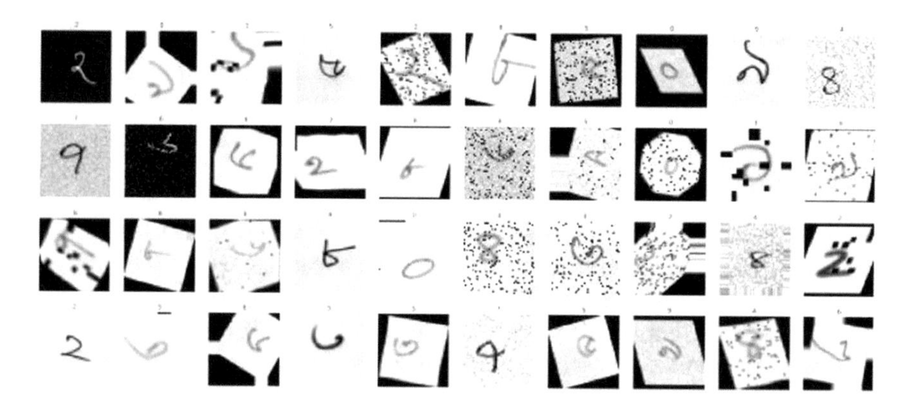

Fig. 5 Augmented images

2.4 CNN Model Architecture and Training Parameters

As we know that setting and tuning hyper-parameters for a neural network is a difficult task, as most of it is dependent upon trial and error methodology. Therefore, after numerous simulations, we designed our model. The parameters of our model include Conv2D, BatchNormalization, MaxPooling2D and Dropout (Fig. 6).

We propose a model with 'Adam' optimizer and we set the learning rate to 0.001. For the first, second and third convolution, we set filter size 32, 64 and 64, respectively. The kernel size is set at (5×5). We set activation function as rectified linear unit (ReLU) with 'same' padding. For the fourth convolution, the filter size is set at 128 and kernel size set at (3×3). We incorporate 'same' padding with ReLU activation. We also add a (2×2) max pooling layer. We then flatten the layer and incorporate

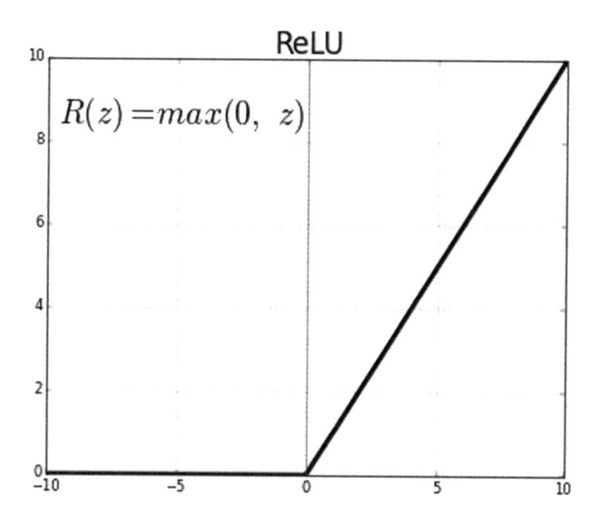

Fig. 6 ReLU activation function

Fig. 7 Softmax function

$$\sigma(Z)_j = \frac{e^{Z_j}}{\sum_{k=1}^{K} e^{Z_k}} \quad For\, j = 1, \ldots, k$$

a dense layer with 1024 units alongside ReLU activation layer with 20% dropout, to avoid over-fitting. At the final output layer of the architecture, we propose to use a softmax activation function with 10 units and then we compile while setting loss = 'categorical_crossentropy' and metrics = ['accuracy'] (Fig. 7).

Now, we set the training parameters and thereby we set the batch_size to 64. Here, we incorporate the function ReduceLROnPlateau which reduces the learning rate as the model learns with ongoing epochs. We use this because models frequently benefit from decreasing the learning rate by a factor of 2–10 once learning stagnates. This callback screens a quantity and if no improvement is seen for a 'patience' number of epochs, the learning rate is decreased.

Next, we fit the model with the following listed hyperparameters:

5000 Training Images (x = X_train[:5000],
y = y_train[:5000]) epochs = 40
Verbose = 1
500 Validation Images ((X_val[:500], y_val[:500]))

3 Result Analysis

Our model performed remarkably well on the NumtaDB dataset even with the added parameters to ensure we deliberately work with a lesser amount of data and the augmented data that we manifested as mentioned above. Each epoch took around $t = 87$ s to complete. For 40 epochs, we attained an accuracy of 98.40% and validation accuracy of 83.20%, while ensuring the model does not get over-fitted in any manner.

The proximity between train and test data is the testimony to the fact of how efficiently our model performed (Table 1; Figs. 8 and 9).

Another set of analysis that we derived is from an experiment we conducted on 20 subjects who are all university students unbiased from gender and age. The survey had four questions and they were asked to answer and rate in the ranks of 0–9, where 0 denotes 'Very Bad' and 9 denotes 'Very Good'. The numbering of the nine digits of the ranks was done in Bengali. The survey had the following questions:

1. How do you like your cellular device? (0–9)
2. How do you like your network service provider? (0–9)
3. How do you like the faculty in your discipline of study? (0–9)
4. How much satisfied are you with the empirical laboratory work that happens at your university? (0–9)

Table 1 Model summary of our deep CNN architecture

Conv2d_1 (Conv2D)	(None, 48, 48, 32)	832
batch_normalization_1 (Batch	(None, 48, 48, 32)	128
max_pooling2d_1 (MaxPooling2	(None, 24, 24, 32)	0
dropout_1 (Dropout)	(None, 24, 24, 32)	0
conv2d_2 (Conv2D)	(None, 24, 24, 64)	51,264
batch_normalization_2 (Batch	(None, 24, 24, 64)	256
max_pooling2d_2 (MaxPooling2	(None, 12, 12, 64)	0
dropout_2 (Dropout)	(None, 12, 12, 64)	0
conv2d_3 (Conv2D)	(None, 12, 12, 64)	102,464
batch_normalization_3 (Batch	(None, 12, 12, 64)	256
max_pooling2d_3 (MaxPooling2	(None, 6, 6, 64)	0
dropout_3 (Dropout)	(None, 6, 6, 64)	0
conv2d_4 (Conv2D)	(None, 6, 6, 128)	73,856
batch_normalization_4 (Batch	(None, 6, 6, 128)	512
max_pooling2d_4 (MaxPooling2	(None, 3, 3, 128)	0
dropout_4 (Dropout)	(None, 3, 3, 128)	0
flatten_1 (Flatten)	(None, 1152)	0
dense_1 (Dense)	(None, 1024)	1,180,672
dropout_5 (Dropout)	(None, 1024)	0
dense_2 (Dense)	(None, 512)	524,800
dense_3 (Dense)	(None, 10)	5130
Total params: 1,940,170 Trainalbe params: 1,939,594 Non-trainable params: 576		

All the participants rated their answers, and we therefore now had the feedback of 20 individuals on these aforementioned topics. We observed that using these numeric data and our algorithm we could perform a sentiment analysis on all the participants and their line of thought with regard to the aforementioned survey. Therefore, we concluded that our empirical results assure us that digitizing the Bengali numerals could also help us analyse a subject's sentiments.

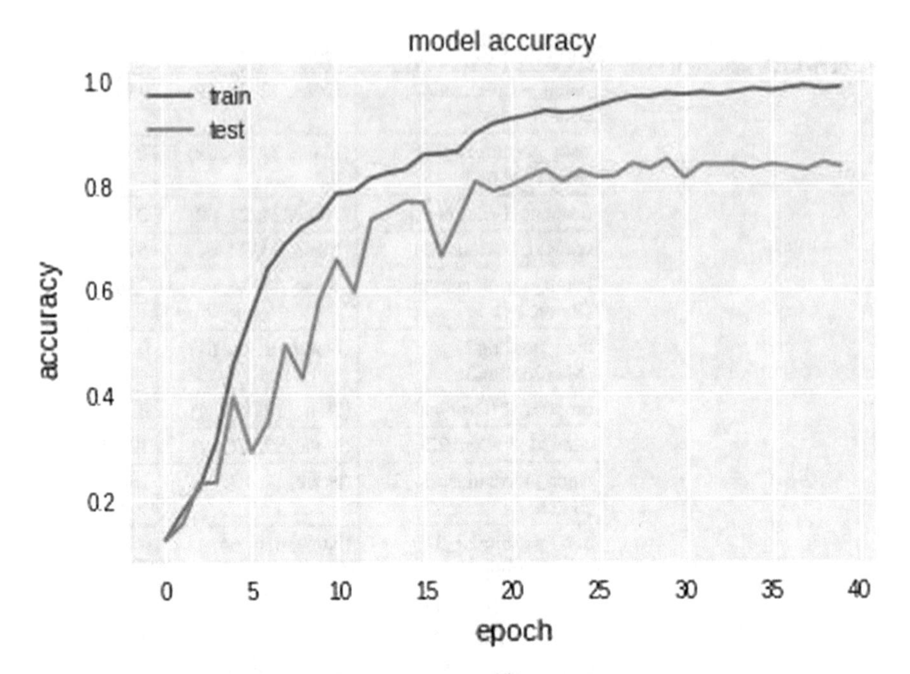

Fig. 8 Accuracy versus epoch plot

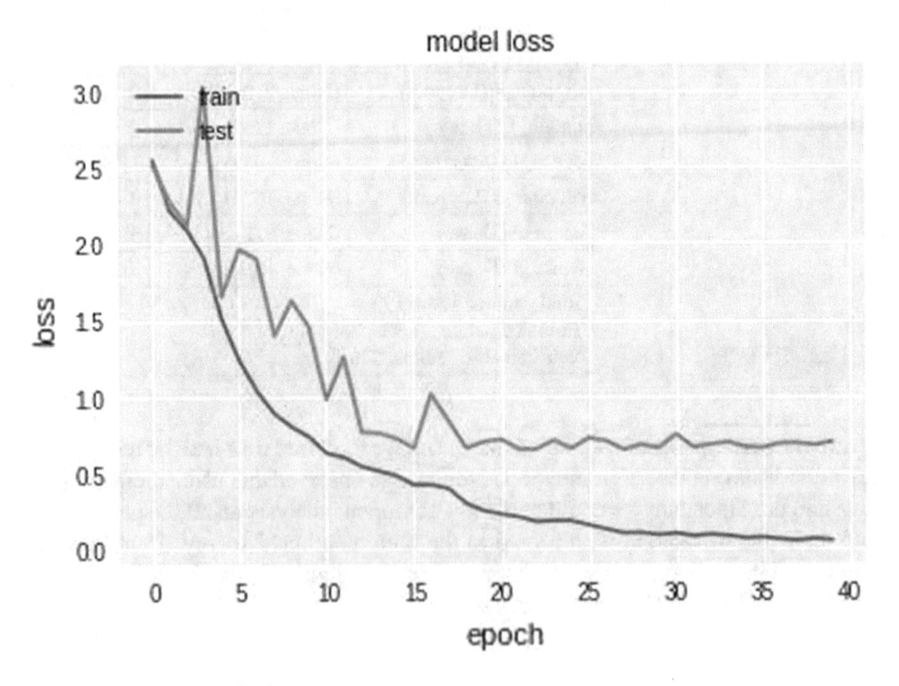

Fig. 9 Loss versus epoch plot

4 Conclusion

In this paper, we methodically worked with the CNN model that we designed and used the same to work with the NumtaDB dataset and therefore achieved a test accuracy of 98.40%. Furthermore, we also performed a sentiment analysis on the survey that we conducted and therefore touched upon a lesser worked periphery of the application of digit recognition from regional texts or scriptures. We further would like to elucidate that this research only provides us with a road map with the area of applicability of language processing and we further aim to ponder upon a model which would classify vernacular alphabets and digits on a sparse dataset using a model similar to what we have used here and without over-fitting the same.

References

1. Pal, U., & Chaudhuri, B. B. (2004). Indian script character recognition: A survey. *Pattern Recognition, 37*, 1887–1899.
2. Bathla, A. K., et al. (2016). Challenges in recognition of Devanagari scripts due to segmentation of handwritten text. In *International Conference on Computing for Sustainable Global Development (INDIACom)*, 2016.
3. Kumar, M., Jindal, M. K., & Sharma, R. K. (2014). Segmentation of isolated and touching characters in offline handwritten Gurmukhi script recognition. *International Journal of Information Technology and Computer Science, 2*, 58–63.
4. Basu, S., Sarkar, R., Das, N., Kundu, M., Nasipuri, M., & Basu, D. K. (2005). Handwritten Bangla digit recognition using classifier combination through ds technique. In *Pattern recognition and machine intelligence* (pp. 236–241). Berlin, Heidelberg: Springer.
5. Pal, U., Chaudhuri, B. B., & Belaid, A. (2006). A complete system for bangla hand-written numeral recognition. *IETE Journal of Research, 52*(1), 27–34.
6. Bhattacharya, U., & Chaudhuri, B. B. (2009). Handwritten numeral databases of Indian scripts and multistage recognition of mixed numerals. *IEEE Transactions on Pattern Analysis and Machine Intelligence, 31*(3), 444–457.
7. Wen, Y., & He, L. (2012). A classifier for Bangla handwritten numeral recognition. *Expert Systems with Applications, 39*(1), 948–953.
8. Maas, A. L., Hannun, A. Y., & Ng, A. Y. (2013). Rectifier nonlinearities improve neural network acoustic models. In *Proceedings of the 30th International Conference on Machine Learning*, Atlanta, Georgia, USA, 2013
9. Rahman, M. M., Akhand, M. A. H., Islam, S., Shill, P. C., & Rahman, M. M. H. (2015). Bangla handwritten character recognition using convolutional neural network. *International Journal of Image, Graphics and Signal Processing, 8*, 42–49.
10. Akhand, M. A. H., Ahmed, M., & Rahman, M. M. H. (2016). Convolution neural network based handwritten Bengali and Bengali-English mixed numeral recognition. *International Journal of Image, Graphics and Signal Processing, 9*, 40.
11. Shopon, M., Mohammed, N., & Abedin, M. A. (2017). Image augmentation by blockyartifact in deep convolutional neural network for handwritten digit recognition. In *2017 IEEE International Conference on Imaging, Vision Pattern Recognition (icIVPR)* (pp. 1–6).
12. Choudhury, A., Rana, H. S., & Bhowmik, T. (2018). Handwritten Bengali numeral recognition using hog based feature extraction algorithm. In *2018 5th International Conference on Signal Processing and Integrated Networks (SPIN)* (pp. 687–690).
13. Sufian, A. BDNet: Bengali handwritten numeral digit recognition based on densely connected convolutional neural networks. Preprint.

Internet of Things-Based Security Model and Solutions for Educational Systems

Ranjit Patnaik, K. Srujan Raju, and K. Sivakrishna

Abstract Today the applications of Internet of things (IoT) are progressing rapidly in variety of domains. This encouraged to develop new applications (e.g., smart grid, smart home, smart cities, wearables, and vehicle networking) advancement as well. The emerging application of IoT is exposed toward security, privacy issues, and its challenges. The main objective of this work is to enhance security in the educational system (ES) using IoT devices. We propose several techniques to avail device identification, authenticate the user, and collect the data from various devices. As the IoT sensors are easily negotiable, it allows unauthorized users/devices that are able to steal and override the data from the cloud. This paper represents a brief summary of IoT security threats and challenges and their classification based on the application domain. The authors identified challenges in security issues in IoT-based educational systems and some probable solutions on security. In this research, the authors propose the incremental Gaussian mixture model (IGMM), blockchain, and EdgeSec as a probable solution for security and machine learning (ML) techniques. In this model, few solutions, like IGMM for authorizing the device, blockchain for the encryption of data during transfer in the information network, ML algorithms for identifying and authorizing devices, and EdgeSec, offer a security profile to collect a huge amount of data about each device in the connected IoT environment. The identified model is used for enhancing security in IoT-based educational systems.

Keywords Machine learning · Deep learning · IoT and its security issues

R. Patnaik (✉) · K. Sivakrishna
Department of Computer Science and Engineering, GIET University, Gunupur, India
e-mail: ranjitpatnaik@giet.edu

K. Srujan Raju
Department of Computer Science and Engineering, CMR Technical Campus, Hyderabad, India
e-mail: ksrujanraju@gmail.com

R. Kumar et al. (eds.), *Multimedia Technologies in the Internet of Things Environment*, Studies in Big Data 79, https://doi.org/10.1007/978-981-15-7965-3_11

1 Introduction

At present, Internet of things is an undetectable part of everyone's day-to-day life. So, it is spread widely in every country from rural to city life. In this paper, we want to highlight the issues and their probable solutions in IoT. Present-day teaching–learning pedagogy no longer confirmed to transform the content of knowledge to the students which they can implement in a real-world scenario. The introduction of Internet of things (IoT) has changed the complete teaching–learning pedagogy to an advanced level, and the machine learning technologies made rich in providing the analysis and future prediction with the help of the learning analytics system (LAS) [1]. The faster architecture of IoT devices capable generates voluminous data, to analyze and predict a model which handy in teaching–learning pedagogy the machine learning techniques made easy. As we know that the data generated by IoT devices are huge in nature, big data, handling these data is difficult without machine learning techniques. The introduction of IoT has changed the computing paradigm. IoT devices mainly consist of sensors, actuators, and microcontrollers, through which it generates voluminous data from its surrounding. These data mainly traveled through a wireless medium, so security plays a vital role to maintain privacy. When these data are analyzed with an accuracy, which able to suggest an appropriate and informal decision. Machine learning is examined as an effective tool for performing analysis and future prediction. This paper discussed an IoT-based education system's application, advantages, and challenges of using machine learning techniques based on data representation. Machine learning consists of future extraction and classification of traditional machine learning. In this paper, the author proposed an IoT-based educational system consisting of input of data from IoT devices directly to the machine learning system where it is classified and future prediction algorithms are implemented that yield the output.

As the IoT devices generate huge data, various types of data are transmitted from source device to the other device in the intranet and Internet. The IoT devices are easily negotiable so there will be a threat of data loss. The data loss leads to the threat to revile the privacy of the user. Hence, the IoT devices must be embedded with security.

In IoT-based educational system, an application of IoT generates a large amount of data and transmits various devices through the network, so the data must be secured by using certain security techniques like EdgeSec [2], machine learning algorithms [3], elliptic curve cryptography [4], blockchain, IGMM, SDN [5], etc.

2 Background

2.1 Internet of Things

IoT consists of sensors, microcontrollers, and actuators. The sensors collect the information from the surrounding. Then, it is transferred to the microcontroller, and the data are sent to the cloud or system for analysis and prediction. Figure 1 shows the architecture of IoT. Data are transferred to the base station through the network or wireless medium. Figure 1 shows the architecture of IoT Technology. IoT is consisting of three layers, each layer is performing a different task. The three layers are (1) application layer, (2) network layer, and (3) physical layer. In Chen et al. [6], the author discussed the three-layer architecture of the IoT environment.

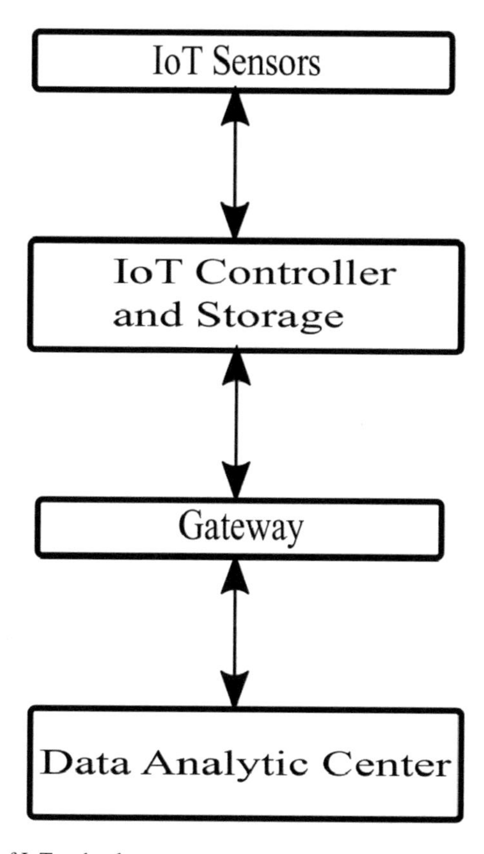

Fig. 1 Architecture of IoT technology

2.1.1 Application Layer

The application layer is the external layer of the IoT environment which provides the user interface or APIs to the users to interact with the devices in the IoT environment.

2.1.2 Network Layer

This layer provides a communication model to communicate among the devices and the base system, and this provides a way to an interface or API.

2.1.3 Physical Layer

This layer describes the physical components like sensors, actuators, storage, controller, etc. Here, the connection between the device is physically attached and configured for the setup of the IoT environment.

2.2 Data Flow in IoT Environment

Figure 2a, b shows the dataflow work on how in the IoT environment using machine learning or deep learning. Figure 2a describes the IoT device request for the neural network algorithm to the machine learning server and then sends the fore part of the neural network. Figure 2b describes key requesting for authentication to the machine learning server, then the expected pattern is provided to the analysis system, and then it gets the result which gives the authentication through the key.

2.3 Learning Analytics (LA)

The learning data and information are collected, based on the collected data or information that the learning pattern is identified and predicted the learning methods. This defines the technique of learning analytics (LA).

2.4 Learning Analysis System (LAS)

The learning analysis system (LAS) is based on the learning analytics (LA) and deep learning, from which an algorithm is proposed which is called the LAS algorithm. This algorithm is used for long short-term memory (LSTM) network purpose.

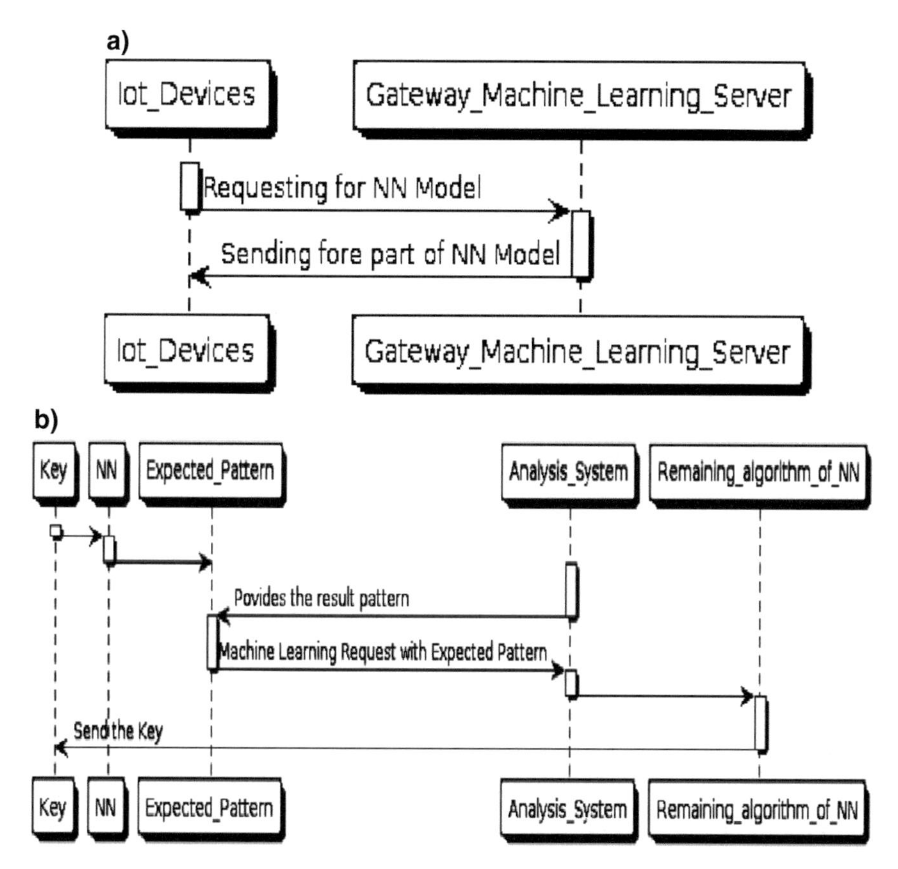

Fig. 2 Data workflow in the IoT environment

2.5 The IoT-Based Educational Model

In this paper, the author proposed an IoT-based educational system consisting of input of data from IoT devices directly to the machine learning system, where it is classified, and future prediction algorithms are implemented that yield the output. Figure 3 shows the base architecture of IoT-based system, which takes the data as input then it is processed in the system using a machine learning technique and gives the output.

2.6 The Architecture of IoT-Based Education System

Ahad et al. [1] demonstrate the architecture of the IoT-based education system and explain the working model where the model takes the input from the IoT devices, and

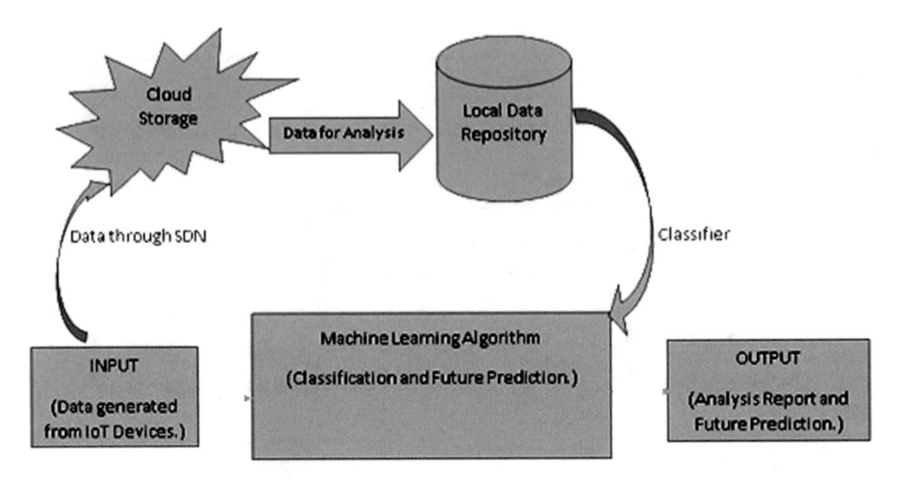

Fig. 3 Proposed architecture of IoT

then the collected information is stored in intermediate storage, i.e., cloud storage and a local data repository system. Data generated by the IoE devices are stored in the cloud storage. Whenever a portion of data is required for analysis, then it is brought into the local data repository and given to the base system to apply machine learning or deep learning algorithm. At last, it generates the report and prediction.

3 The Proposed Architecture of Base IoT Environment

Figure 3 shows that the proposed model that takes the input from the IoE devices, and then the collected information is stored into intermediate storage, i.e., cloud storage and a local data repository. Data generated by the IoE devices are stored in the cloud storage. Whenever a portion of data required for analysis, then it is brought into the local data repository and given to the base system to apply machine learning or deep learning algorithm. At last, it generates the report and prediction.

3.1 The Proposed Architecture of IoT-Based Educational System

Figure 4 shows the different phases of the proposed architecture of IoT-based educational system, herein describes in detail the various work performed in the different phases of the IoT-based educational system.

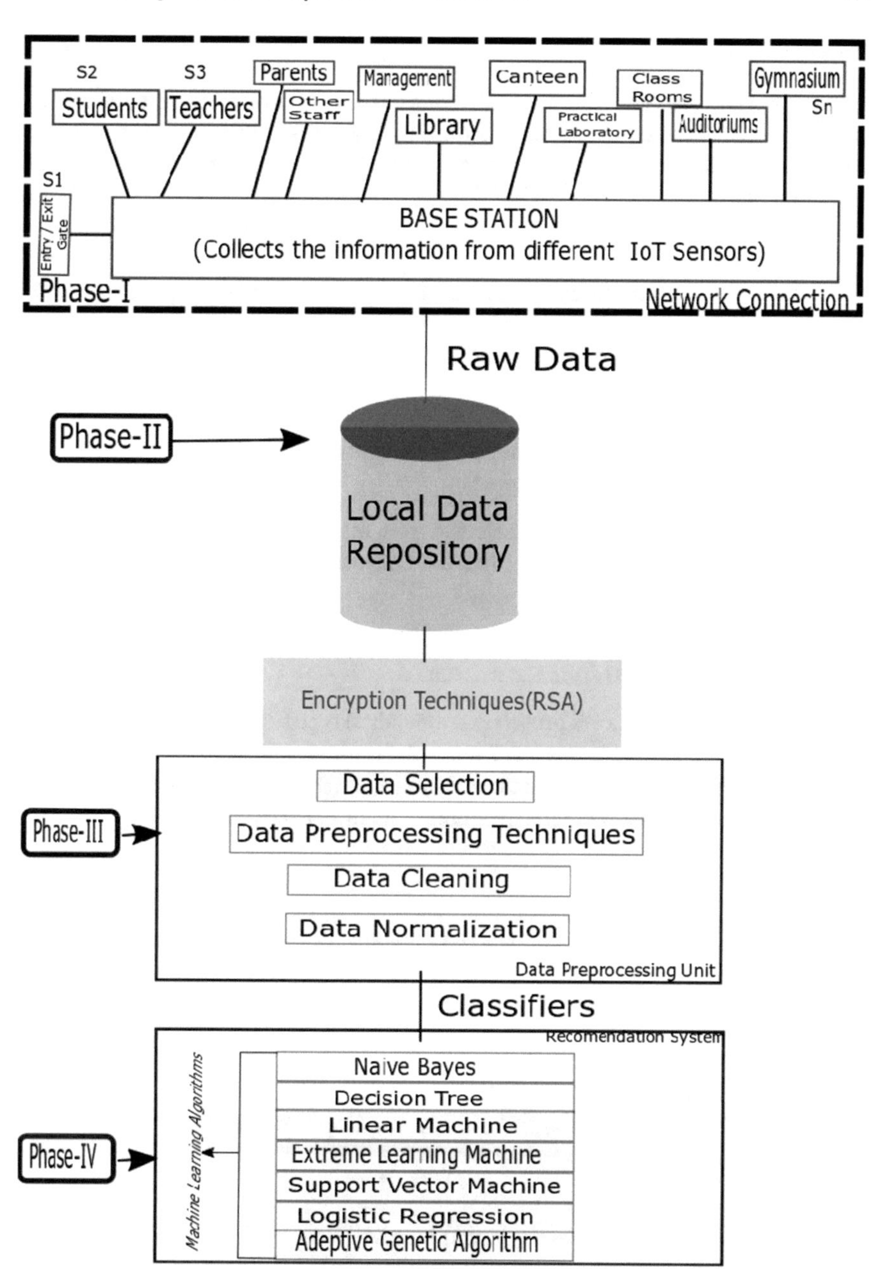

Fig. 4 Proposed architecture of IoE-based educational system

Phase I

The IoT sensors are collecting the information from the entities and their surroundings, which transmitted to the base station. In that station, the data are stacked. In this phase, the data are in the form of raw data. The raw data are having noise and unwanted data.

Phase II

The stacked raw data are stored in the local data repository for further processing. The stored data are encrypted by using the encryption algorithm before it is transmitted to the base system for the further process through the information network.

Phase III

The raw data from the base station are undergoing the following stages in data preprocessing unit:

Data selection: Here, the dataset is selected from the sensors of the IoT environment.

Data preprocessing techniques (data are converted into a standard format from the varied format.)

Data cleaning: Here, the raw data are separated from the noise and unwanted data from the dataset collected from the raw data.

Data normalization (Data are bringing to a standard type from heterogeneous types.)

After completion of above steps, the raw data from the sensors are converted into the standard data format. The standard data are processed for encryption by using cryptography techniques, which will be able to identify and access only by authorized personals.

Phase IV

The preprocessed data are implemented by various classifier algorithms of machine learning techniques. A proposed teaching–learning approach is defined using learning analysis on the basis of techniques in machine learning.

4 IoT Threats and Attacks Based on the Taxonomy of the IoT

In this paper, authors discussed the probable solution for security in IoT. IoT devices are interconnected through the network and through the network data will be transmitted, and this causes an area of threats. Blockchain is one of the techniques which can provide security to the data transmitted in the network.

4.1 Labeling of Security Issues

IoT technology contains several devices and equipment from huge high-end servers to small processing chips rooted devices. It addresses security issues at different levels.

In Table 1, the tick (✓) mark indicates that the authors have argued different types of IoT security issues and they had explained about it. The systematic literature survey is conducted from 2003 to 2018. The blank space indicates that the authors have not argued about the issues.

In Fig. 4, the *x*-axis describes the years from 2003 to 2018, and the *y*-axis describes the referred papers on the various IoT security issues. This figure describes that during the years researchers have emphasized the various types of issues on security in IoT. Each author referred to the issues each year.

4.2 Security Issues in Physical Level

The issues in the data link and physical layer of communication are associated with this kind of security issue category. The issues are discussed below are classified in this type of security issues:

4.2.1 Jamming Adversaries

This kind of attack is done by decreasing of network that sent frequency signal without following the any protocols. The radio interference largely affects the network operation like sending and receiving of data in IoT.

4.2.2 Insecure Initialization

To ensure a proper and secure network service in IoT, we must initialize and configure IoT devices in the physical layer without deviating secrecy and obstacle to the network.

4.2.3 Low-Level Sybil and Spoofing Attacks

The random forged MAC values are used by Sybil node to deceptive as a masking device while aiming at the diminution of networker sources [7, 8].

4.2.4 Insecure Physical Interface

Debugging exploits to negotiated nodes in the network by the poor physical security with the help of physical interface and tools [9].

4.2.5 Sleep Deprivation Attack

IoT devices are designed to consume low power so these are getting into sleep mode. The attackers are trying to not enter the device into sleep mode. It is known as sleep deprivation [10].

4.3 Security Issues in Middle Level

The communication, session management, and routing taking place at "network and transport layers of IoT" are concerned in middle-level security are explained as:

4.3.1 Duplication or Repetition Attacks Due to Fragmentation

Packets and fragments reconstructed may raise to overflow of the buffer and rebooting of devices at 6LoWPAN [11]. Barricading the accessing of authentic packets due to the redundant fragments sent by a malevolent node to packet reconstitute is affected [12].

4.3.2 Insecure Neighbor Discovery

Neighbor nodes are discovered, prior to router discovery and address resolution for transmission of data [13].

4.3.3 Buffer Reservation Attack

Nodes are accessing the buffer to send or receive packets and reserving buffer for regathering of storing of packets, by sending broken packets it may be exploited by an attacker [14].

4.3.4 Routing Protocol for Low-Power Networks Routing Attack

In IPv6 routing protocol for low-power networks (RPL) may have vulnerably elicited in network because of negotiated nodes [15].

4.3.5 Sinkhole and Wormhole Attacks

In this attack, malicious activity in the network can be raised due to the packets from the attacker node on responding to the routing requests [16, 17]. The operations of 6LoWPAN may be further declined due to the attacks on the network in the channel created among the nodes so that data reach at a node due to wormhole attack [18–20].

4.3.6 Sybil Attacks on the Middle Layer

Like Sybil attacks on lower-level layers, the nodes become a Sybil can be generalized to decrease network performance and volatile privacy of the data. As the Sybil nodes are imitating the characteristics of other nodes in a network, it leads to broadcasting malware, spamming, or phishing attacks [20, 21].

4.3.7 Authentication and Secure Communication

Any fault in network layer security can be exposed to the network to a numerous vulnerability [22–24]. It is ensured that the use of cryptographic mechanisms secure data communication over IoT devices [25, 26].

4.3.8 Transport Level End-To-End Security

Data are transmitted from node to node by the services of transport level. It must be ensured that it should be transmitted in a secure way and authentic node [27, 28]. A compact authentication mechanism is required to provide a secured message communication in the form encrypted without revealing privacy [29, 30].

4.3.9 Session Establishment and Resumption

At the transport layer, the fake messages in the session are taken over, which can be resulted in denial of service [31–36].

4.3.10 Privacy Desecration on Cloud-Based IoT

Encroach upon location and individuality privacy is exposed to delay-tolerant network-based IoT or cloud by different attacks [15].

4.4 Security Issues in Logical Level

The issues at the logical level are provided through the applications executing on IoT as follows:

4.4.1 Constrained Application Protocolsecurity Using the Internet

Application layer is consisting of various types of applications for a different purpose, which is sometimes vulnerable. It leads to the attack by the unauthorized attacker [32–34].

4.4.2 Insecure Interfaces

The IoT service interfaces used in cloud, Web, and mobile are responsible for susceptible to several attacks, which causes loss of privacy of the data [9].

4.4.3 Insecure of Software/Firmware

The software/firmware may cause several susceptibilities which make IoT insecure [9].

4.4.4 Middleware Security

The middleware of the IoT is developed to reduce latency in communication among different components of the IoT, which are to be secured enough as the provider of services [35, 36].

The above-cited issues in IoT are shown in Fig. 5, which describes the issues in different layers in the IoT environment.

Figure 6 shows the papers referred to the issues in the IoT in different years. Here, the x-axis and y-axis imply to the year of publication and referred papers on the issues in IoT. Between the years 2010 and 2012, more issues are discussed than the other years. The above figure is derived from Table 1.

5 Results and Discussion

During the implementation of the IoT and ML techniques in the educational system, the authors used the dataset of the students of the 4 years of B.tech. Each batch consists of 11 numbers of branches and each year has 2 semesters. The dataset consists of

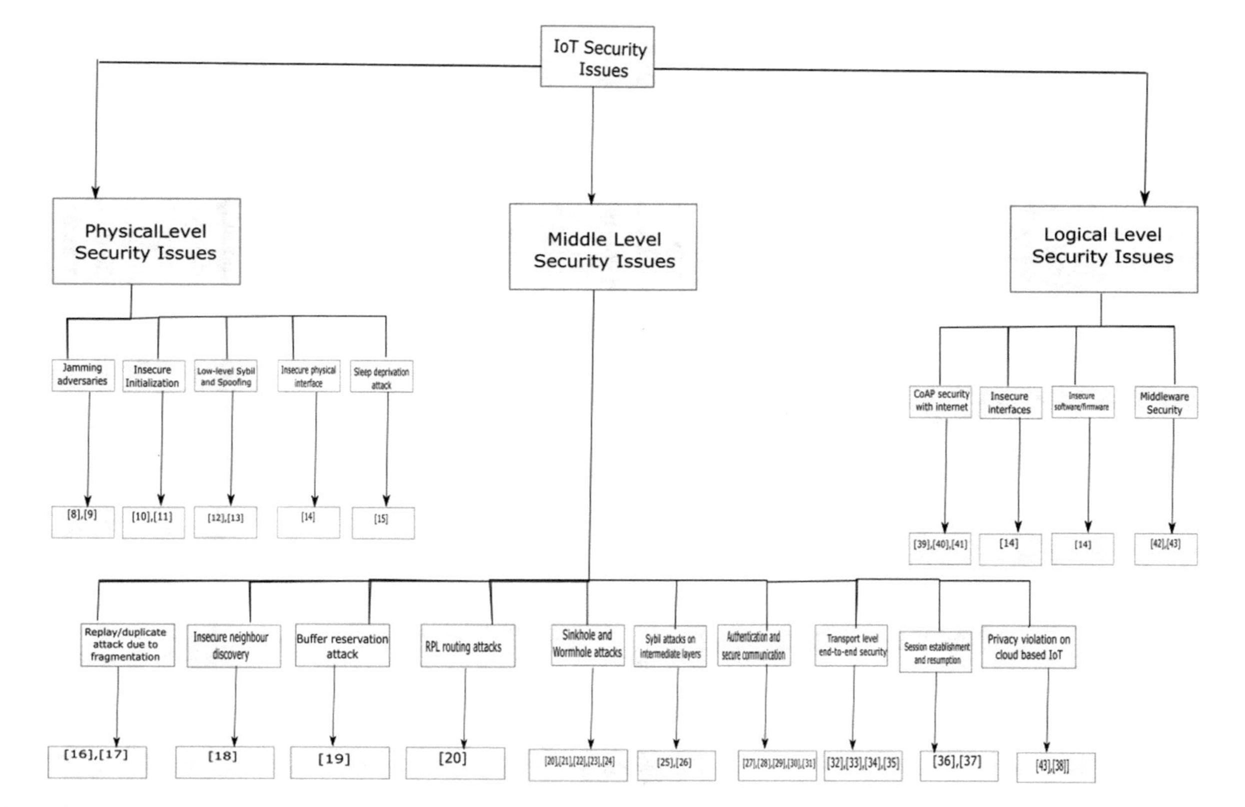

Fig. 5 IoT security issues and related publications

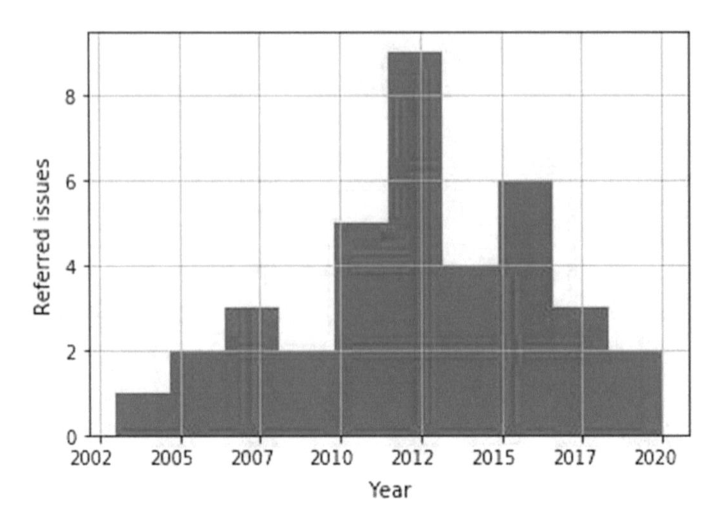

Fig. 6 IoT security issues referred to in publication year-wise

33 columns and 3950 rows. From the dataset, we implemented the following ML algorithms. In the dataset, we considered the performance in the class and examination of the students and then calculated the grades of the individual students. From the calculation of the grades, the distribution of the grades is derived. During the implementation of the dataset, we have used the high configuration computing server which is given in Table 2.

In this research paper, the researchers used machine learning algorithms for predicting the data analysis. Here, the few algorithms of machine learning on the dataset of the students to predict the grades are implemented. Figure 7 shows the final grade distribution. The above figure is derived based on the distribution of grades in Table 3. Herein it describes the initially we considered the 50% as a train set which yields the accuracy of 70%, whereas it considered 25% train set and 75% test set it gives around 90% accuracy.

Table 4 describes the mean absolute error (MAE) and the root mean square error from the various ML algorithms that are implemented over the dataset of the students. The grades of the students are predicted by using several ML algorithms based on regression which gave the above MAE and RMSE with the help of which we can predict the best model for the grade prediction of the student. While comparing the different algorithms we found that the mean absolute error (MAE) and the root mean square error (RMSE) are minimum for **gradient boosted** algorithm hence the accuracy of the **gradient boosted**is maximum and it has the minimum deviation so the best algorithm for grade prediction in this dataset is **gradient boosted**.

Table 1 IoT security issues and their referenced publications

Year	Reference No.	IoT Issues in Security																		
		Security issues in Physical level					Security issues in Middle level										Security issues in Logical-level			
		1	2	3	4	5	1	2	3	4	5	6	7	8	9	10	1	2	3	4
		Jamming adversaries	**Low-level Sybil and spoofing attacks**	**Insecure initialization and configuration**	**Insecure physical interface**	**Sleep deprivation attack**	**Replay or duplication attacks due to fragmentation**	**Insecure neighbor discovery**	**Buffer reservation attack**	RPL routing attack	**Sinkhole and wormhole attacks**	Sybil	**Authentication and secure communication**	Transport level end-to-end security	**Session establishment and resumption**	**Privacy violation on cloud-based IoT**	**CoAP security with internet**	**Insecure interfaces**	**Insecure software/firmware**	**Middleware security**
2003	[15]	√																		
2005	[28]										√									
2005	[14]	√																		
2007	[18]		√								√	√								
2008	[22]						√													
	[29]										√									
2009	[19]		√									√								
	[24]							√												
2010	[33]												√							
2011	[21]					√														
	[25]								√											
	[32]												√							
	[36]													√						
2012	[3]													√						
	[37]																√			
	[38]																√			
	[39]																			√
	[25]									√	√									
2013	[40]											√								
	[34]												√							
	[36]													√						
	[37]																√			
2014	[41]																			√
	[40]											√								
	[31]												√							
	[16]			√																
2015	[40]											√								
2016	[20]				√													√	√	
	[27]										√									
	[30]										√									
	[42]														√					
	[43]														√					
2017	[44]															√				
	[41]															√				
2019	[6]																			√

Table 2 Server configuration

S. No.	Component name	Configuration
1	CPU	Quad-core
2	RAM	32 GB
3	Storage	1 TB
4	GPU-RAM	12 GB
5	Python with PyTorch	Must installed

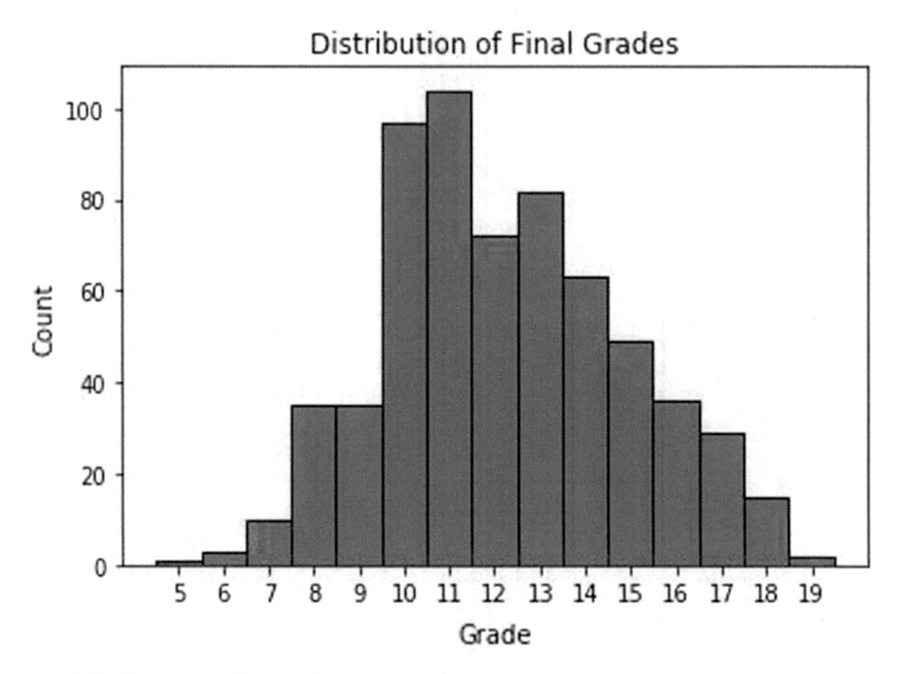

Fig. 7 Final grade distribution of the students

Table 3 Distribution of grades

Count	633.000000
Mean	12.205371
Std	2.656882
Min	5.000000
25%	10.000000
50%	12.000000
75%	14.000000
Max	19.000000

Table 4 MAE and RMSE estimated from the implementation of the ML algorithms

List of algorithms	MAE	RMSE
Linear regression	1.8859	2.27395
Elastic net regression	2.09282	2.53656
Random forest	1.85529	2.32467
SVM	1.90262	2.30825
Gradient boosted	1.81272	2.22737
Baseline	2.1761	2.67765

5.1 *Error Estimation*

During implementation, there are different types of error estimation for the distribution of grades for the different students all these errors defined based on the total number of samples.

5.1.1 Mean Absolute Error (MAE)

Mathematically, MAE is obtained using

$$\text{MAE} = \frac{1}{n}\sum_{i=1}^{n}\left(\left|Y_i' - Y_i\right|\right) \tag{1}$$

Where Y_i' refers to the calculated output, while Y_i is for the expected value.

5.1.2 Mean Magnitude of the Relative Error (MMRE)

Mathematically, MMRE is obtained as

$$\text{MMRE} = \frac{1}{n}\sum_{i=1}^{n}\frac{\left|Y_i' - Y_i\right|}{Y_i} \tag{2}$$

where Y_i' signifies the estimated output, while Y_i defines the expected value. To deal with overflow conditions [when denominator becomes Zero ($Y_i = 0$)], here added 0.01 in the denominator. Thus, MMRE is obtained as,

$$\text{MMRE} = \frac{1}{n}\sum_{i=1}^{n}\frac{\left|Y_i' - Y_i\right|}{Y_i + 0.01} \tag{3}$$

5.1.3 Standard Error of the Mean (SEM)

Mathematically, SEM is presented below

$$\text{SEM} = \frac{\sigma}{\sqrt{N}} \tag{4}$$

Where σ and N present the standard deviation and the total number of samples, respectively.

Figure 8 describes the mean absolute error values in the various machine learning algorithms.

Figure 9 describes the root mean square error values in the various machine learning algorithms.

5.2 Research Questions

During the implementation of the above algorithms for prediction of the performance of the students in the IoT-based educational system, these are the following research questions addressed in this paper.

- **RQ 1** *How a device is identified and authorized for sending and receiving data?*
- **RQ 2** *How to protect the data from unauthorized and malicious access?*
- **RQ 3** *How the devices are identified in the interoperable network?*
- **RQ 4** *How to define a security profile for the IoT devices?*

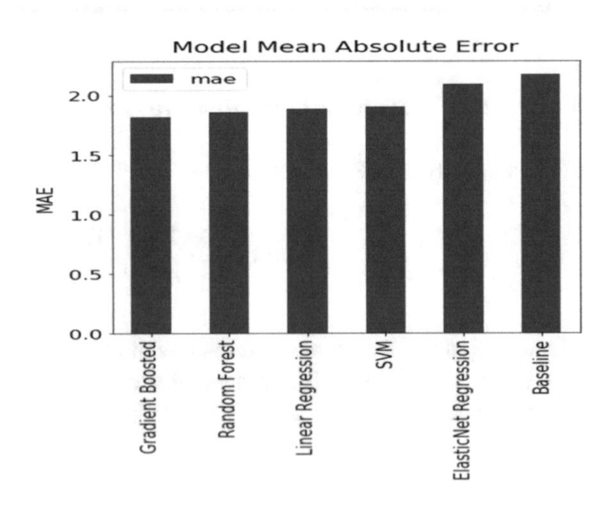

Fig. 8 Model for mean absolute error (MAE)

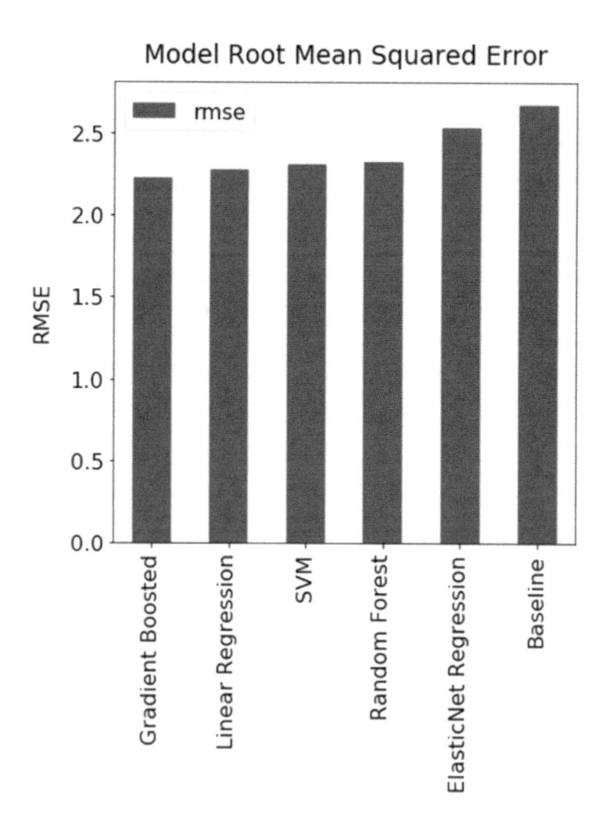

Fig. 9 Model for root mean absolute error (RMSE)

5.2.1 The Probable Solution of IoT Issues

In this process, the IoT device which needs to be specified, access for communication or devices which receives the key from the leader of group is requested to transmit some parameter values like MAC address, received signal strength, and no. of packets the device is able to receive during the predetermined time interval. When the IoT device receives this request message, it replies to the authorized receiver with the corresponding message. The comparison between the received parameter values from the IoT device and the parameter values of the corresponding IoE device which are detected by the receiver itself is made. Then, incremental Gaussian mixture model (IGMM) is applied to verify whether the IoE device requesting the key is belonging to the group or not. In the comparison process if the result is positive and is identified to be in the proximity of the group, then the IoE device is shared with the key by the group leader. Once the IoE device is identified as an authorized device, then the communication can be made with any other authorized IoE devices in the proximity.

RQ 1: How a device is identified and authorized for sending and receiving data?

5.2.2 Incremental Gaussian Mixture Model (IGMM)

IGMM is a probability-density-based data stream clustering approach. The newly arrived data are added into the memory, and the approach incrementally updates the density of newly added data and previously estimated the density of data. This approach is mainly used for online data streaming. The incremental Gaussian mixture model is based on incremental Gaussian mixture network (IGMN). The IGMN is capable of learning from data streams in a single pass by improving its model after analyzing each data point and discarding it thereafter. Nevertheless, it suffers from the scalability point of view, due to its the asymptotic time complexity of $O(NKD^3)$ for N data points, K Gaussian components, and D dimensions, rendering it inadequate for high-dimensional data.

Here, IGMM is used to learn the device's flow in an incremental and unsupervised manner. The probability compactness of the devices into the network $p(a)$ can be demonstrated by a direct mixture of element compactness, i.e., group, $p(a|b)$ equivalent to self-governing probabilistic procedures, in the form Eq. 5.

$$P(a) = \sum_{b=1}^{n} P(a|b)P(b) \tag{5}$$

where $p(b)$ is the mixing parameters like MAC address, received signal strength number of packets arrived in a specified time interval "t" of the device x. It is known that in Eq. 6,

$$\sum_{b=1}^{n} P(b) \quad \text{where } 0 \leq P(b) \leq 1 \tag{6}$$

Element compactness $P(a|b)$ is standardized as in Eq. 7.

$$\int P(a|b)\mathrm{d}a = 1 \tag{7}$$

IGMM utilizes a base probability basis to perceive a vector "a" as having a place with an element to be combined. For every approaching device, the procedure checks whether it insignificantly fit any group. A device "a" is not perceived as having a place with a group "b" if its likelihood $P(a|b)$ is lower than a recently indicated least probability (or curiosity) threshold. Figure 10 is constructed by using the Python open-source tool as a UML perspective (sequence diagram). The list of actions is carried out and discussed below.

Steps

1. The device is seeking access to the data from the other devices, and then it requires the key which will be provided by the group leader.

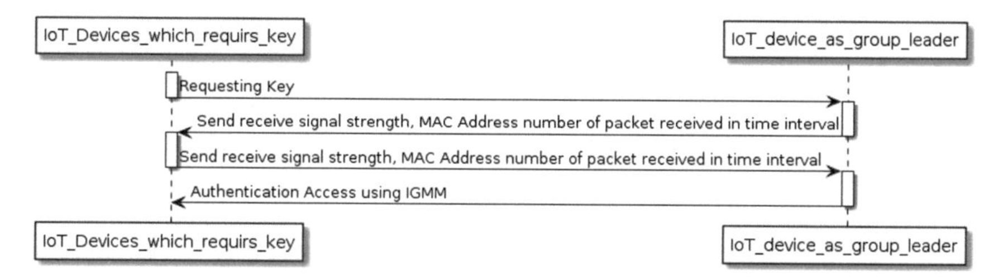

Fig. 10 Implementation IGMM for device authentication model

2. The leader device sends a signal strength and MAC address for the verification of the device which is seeking access.
3. The requesting device again sends the signal strength and MAC address to the group leader.
4. Group leader verifies the details from the information stored in the base system by implementing an incremental Gaussian mixture model (IGMM) which described above Eqs. 5, 6, 7.

By using the above models, the security of the data can be enhanced in IoT and provide authentication to devices.

RQ 2: How to protect the data from unauthorized and malicious access?

5.2.3 Blockchain

A blockchain is a technique that uses the data structure of distributed configuration that can be shared among the components of the network. The blockchain consists of a distributed ledger which is having all transactions within the system, enhanced with cryptography and carried out through the peer-to-peer nodes. This is an innovative and effective solution for the security challenges of centralized tracking, monitoring, and security of data because of its decentralization computing nature. Figure 11 shows the architecture of blockchain. The structure of blockchain is represented by

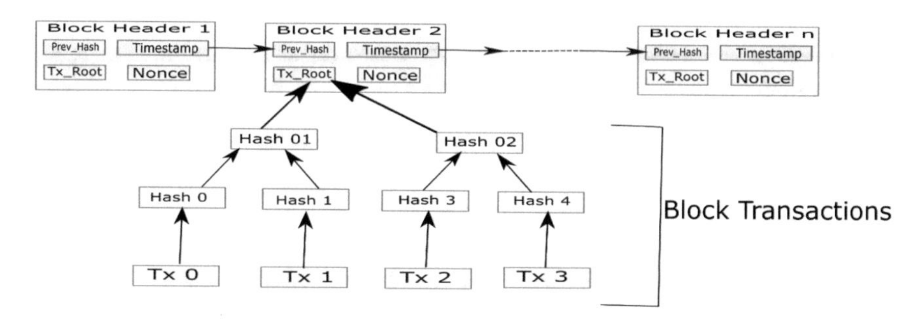

Fig. 11 Architecture of blockchain

a list of blocks with transactions in a particular order. This list stored in the form of flat files (.txt format) is the simple database. Blockchain has two primary data structures: 1. pointer and 2. linked list. The blockchain uses the concepts of unspent transaction output (UTXO) model. The blockchain can able to provide the security in the network. The UTXO model mainly preserves the unspent electronic money in the account of the authorized customer.

In [17], the authors enlighten the blockchain and steganography for securing the data in the IoT-based network. Steganography is the protocol of hiding the content in an encrypted format such a way that if it is the first so it is difficult to identify without decryption. In [45], it described that the elliptical curve version of the blockchain can control the authentication and access control.

RQ 3: How the devices are identified in the interoperable network?

5.2.4 Machine Learning (ML) and Deep Learning (DL)

In [2], the authors discussed the use of machine learning and deep learning algorithms like regression, classification, and density estimation. The various ML and DL algorithms provide the intelligence to the IoT network and devices to tackle the vulnerability, then provide the security. There are multiple algorithms in ML and DL which offer protection in the different levels of IoT. Here, the authors have illustrated some exclusive characteristics of IoT, which lead to challenges in security while deploying IoT. Machine learning provides a technique of learning from past records or data. ML is used by Google to analyze threats against mobile endpoints and applications which are running on Android. To sort and classify data stored in its cloud storage service, Amazon has launched a service called Macie that also from ML. Deep learning is a new variant of machine learning which is a self-service version of machine learning for classification and prediction tasks in innovative IoT applications. Table 5 shows the algorithms in machine learning which are the probable implementation of security of the data in IoT.

Device Identification

The IoT devices are classified into a labeled network, whose information is fed to the system as train data for the machine learning model. It is used to evaluate the classifier. The ML algorithm can accurately identify the IoT devices. Meidan et al. [39] show that the authors discussed the techniques of accurate identification of devices in the IoT networks based on network traffic analysis. Here, authors acquired network traffic data from various IoT devices or sensors and some non-IoT devices, like PC, laptop, smartphone, etc., then analyzed and fed to train the model by using machine learning algorithms for classification of IoT devices to calculate the optimal size of a sequence of data for the selected device, which is to be identified.

Table 5 Machine learning techniques used in IoT security

List of algorithms for IoT security purpose	Description
Gradient booster	It is a machine learning technique for regression and classification problems, which produces a prediction model in the form of an ensemble of weak prediction models, typically decision trees. It optimizes the model according to MSE. in this algorithm since we can't remove outliers, it's better to optimize the mean absolute error (MAE) also called the L1 loss or cost.
Random forest	It is an ensemble technique where a collection of the decision trees. Rather than just simply averaging the prediction of trees this is an ensemble technique that holds a collection of decision tree induced.
Linear regression	Linear regression is a supervised learning algorithm that is used in continuous data. The core idea is to obtain a line that best fits the data. The best fit line is the one for which total prediction error (all data points) are as small as possible. Error is the distance between the point to the regression line

(continued)

Table 5 (continued)

List of algorithms for IoT security purpose	Description
Support vector machine (SVM)	[illegible]
Elastic net regression	[illegible]
Naïve–Bayes	[illegible]

(continued)

Table 5 (continued)

List of algorithms for IoT security purpose	Description
K-nearest neighbor	[illegible]
K-mean algorithm	[illegible]
Recurrent neural network	[illegible]

(continued)

Table 5 (continued)

List of algorithms for IoT security purpose	Description
Q-learning	[illegible]

Table 6 Machine learning techniques used in IoT security

Challenges and issues of IoT security	Machine learning algorithms (probable implementation)
Device identification	[39], a classification algorithm is implemented on the network traffic
Authentication and access control	[45], classification and random forest algorithm along with the elliptical curve of blockchain
Malware and malicious access control	[6], the classification and regression algorithms are used to identify malicious access by unauthorized access

Authentication and Access Control

Kumar et al. [45] show that the authors discussed the use of the elliptical curve of blockchain and the classification, and random forest algorithm is used to check the unauthorized access of the data in IoT-based educational system and put a check on the illegal access on the devices.

Malware and Malicious Access Control

Chen et al. [45] show that the authors discussed the protection of malware and malicious access control by using regression algorithms of machine learning.

Table 5 shows the detailed machine learning algorithms, and each algorithm is used in the appropriate analysis of the data in the educational system.

Table 6 describes the various security issues and probable solution using machine learning algorithms.

5.3 *Deep Learning*

Meidan et al. [5] describe the deep learning algorithms that are used for the detection of an unauthorized device in IoT.

$$F_\beta = \left(1+\beta^2\right).\frac{\text{precision}\cdot\text{recall}}{\beta^2\cdot\text{precision}+\text{recall}} \tag{8}$$

Equation 8, is describing the way, deployment of the methodology and interested to implement an IoT security policy strictly can choose a $\beta < 1$ to put more emphasis on precision than on recall. Thus, fewer unauthorized IoTs connected to the organizational network would be incorrectly identified as whitelisted. Other organizations, wishing to reduce false alarms concerning authorized IoTs, could use a $\beta > 1$.

During the preprocessing stage, the intrusion integration detection estimates meta-features by using the probability distribution concept, and it is represented as below Eq. (9)

$$fp = \{P(B0), P(B1), \ldots, P(B7)\} \quad (9)$$

In the above equation, the parameter *P*(*Bi*) is denoted as the probability of each byte "1" which is observed in the ith byte position, which is represented in Eq. (9).

$$f = L(f0) \quad (10)$$

In Eq. (10), the logical mapping function is defined L: R8 › → R8, i.e., represented as actually if *P*(*Bi*) which is greater than a half and the measurement of the probability is mapped to either 1 or 0.

In the machine learning algorithm, the set of features represented as the tuples is required to input and each record again is represented as a data vector *dv* from f. Apart from this, the meta-features set can be illustrated as a feature vector *fv* at a time instance *n*, generated as,

$$fv(n) = dv(n) \oplus dv(n-1) \quad (11)$$

The operator $\oplus$ used to every place of bits in the vector.

5.4 Proposed Model for Deep Belief Layer

In this section, we have developed our proposed deep learning model which uses the concept of supervised training and binary classification technique. These two approaches used to detect malicious code. During the training phase, if the model deep neural network (DNN) identifies unauthorized anomaly, then it stores details about the record of the filtered features to the cache as feedback. With the help of this feedback, the deep neural network model is retraining. In this way, we enriched the features extracted and labeled the detection system. It may happen that, the features which we have extracted may not be sufficient, then feedback is again sent to the data collection phase and another module, i.e., transmission. This transmission module is responsible for retraining the model.

In order to construct the feed-forward deep neural network (DNN), we have used the deep belief network as the benchmark for the perceptual learning model. Figure 12 is connected to each other layers in an undirected manner. Every layer consists of n number of neural nodes but deep neural network is another kind of feed-forward neural network having different layers. The above model is created for the pre-trained purpose with the help of the unsupervised (US) learning mechanism. The term wi is used for weights for all the hidden layers of the deep belief network model. This model obtains the weights by using unsupervised training. During the training phase, those parameters generated are only means for allocating the initial set of weights.

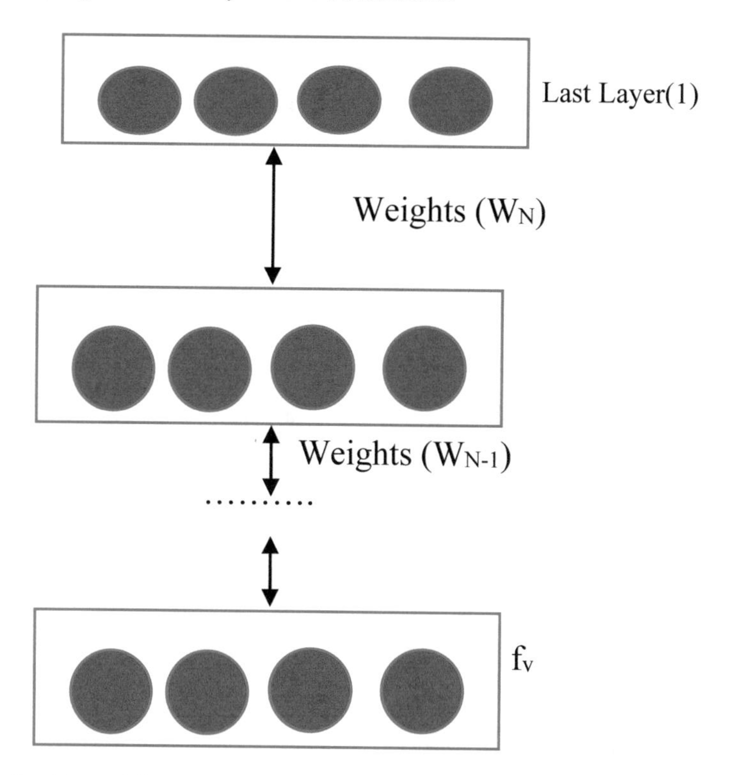

Fig. 12 Feed-forward deep neural network

5.5 Proposed Model for Deep Learning-Based Binary Classification Model

In Fig. 13, the above-mentioned model is for the deep learning-based binary classification layer and the information about the layer is to transform into a deep neural network. The bottom layer is designated which is called a bottom-up supervised learning approach. For the period of the learning process, every node in a deep neural network layer is assigned with some weights as a parameter and influenced by using the novel gradient descent technique.

5.6 Proposed Model for Deep Learning-Based Intrusion Detection System

In this section, we have developed a deep learning-based model for an intrusion detection system which consists of one layer which is the input parameter and the three hidden perceptual layers and obtains one output layer, which is called one

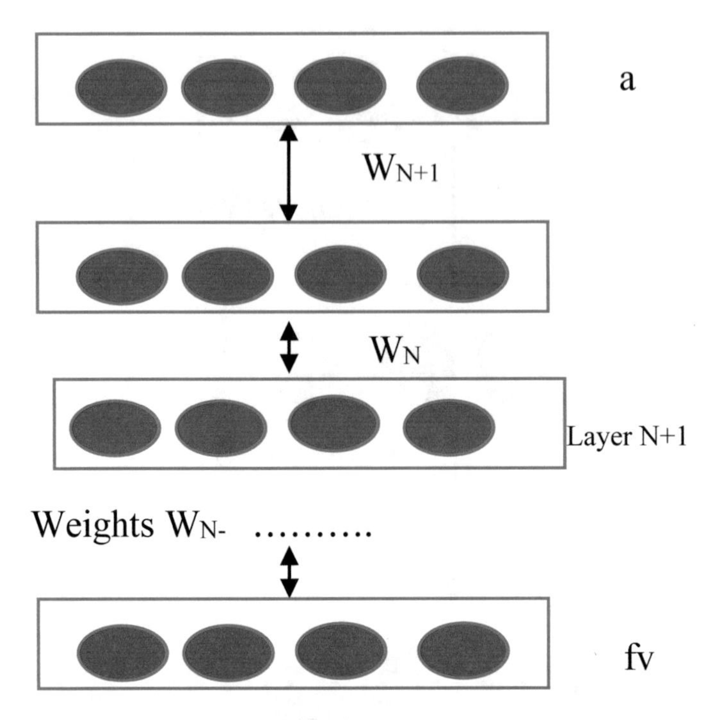

Fig. 13 Deep learning-based binary classification

type of a binary classifier layer. Figure 14 shows the proposed model for the deep learning-based intrusion detection system.

All the input features which are going to feed the model are called the tuples. In this model, we have supplied 56 nodes which are called as primary as well as

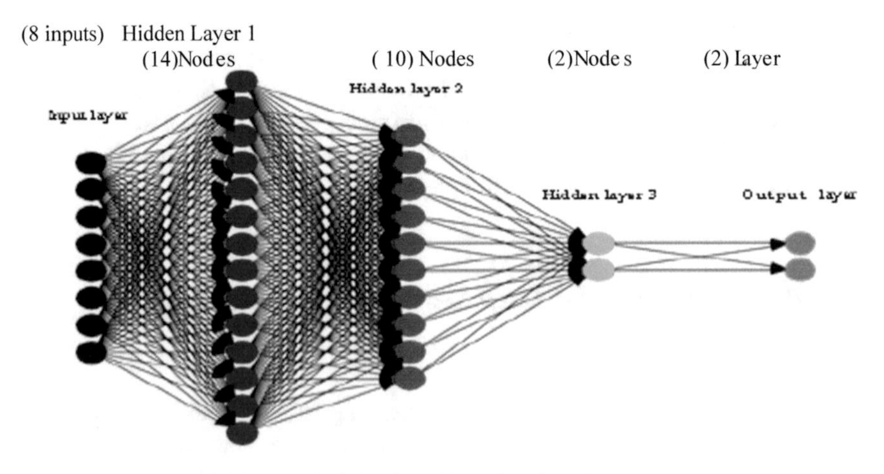

Fig. 14 Proposed model for deep learning-based intrusion detection system

secondary features. During the supervised training phase, the a is fed to the deep neural network. The input passes through from the first hidden encode layer and then scanned (let us say M). Similarly, M will input and generate N features which will again feed to the third layer. The third layer received N and filters the two features and finally, the result is passed to the output layer which represents the classification as malicious and controlled it. The main role of the output layer does not filter any output but also inserts the previous (third layer) and gives the final result. Similarly, all the encoder layers supply the data and generate numerical value. These values are normalized as 0 and 1.

$$\text{Network traffic}\begin{cases} 0 = \text{stands for benign traffic} \\ 1 = \text{stands for malign network} \end{cases} \quad (12)$$

RQ 4: How to define a security profile for the IoT devices?

5.7 *EdgeSec Technique*

In [46], the authors proposed a technique of securing data in the IoT environment is called EdgeSec: Security service in layers of edge, which includes the security options of edge device in the IoT environment. The IoT devices were moreover similar to the edge devices. Edges contain several components like security profile management (SPM), protocol mapping (PM), security analysis module (SAM), protocol mapping (PM), user interface (UI), and interface manager (IM). The IoT devices are registered into the EdgeSec module by SPM, where a security profile is created. In this paper, the enhancement of security in IoT is discussed (Fig. 15).

6 Conclusion and Future Scope

In this paper, we discussed the most imperative aspect of the security issue of the IoT environment with highlighted further research on this area of security in IoT as a survey. In our explore is the overall architecture of IoT and the overall architecture of security in IoT and followed by, the security issues connected to interconnected and operated of different kinds of IoT devices. We also performed a meticulous inquiry of issues on security in the IoT system and the limitations of energy, a lightweight protocol of cryptographic and resources. Furthermore, we cited the circumstances of real life wherein lacking security could pose various threats in IoT. Here, we proposed techniques for the enhancement of security in IoT-based educational systems. The IGMM authenticates the devices from malicious attack. Machine learning and deep learning algorithms provide the identity of devices, authentication, and access control. EdgeSec is the inherited security profiling from edge technology

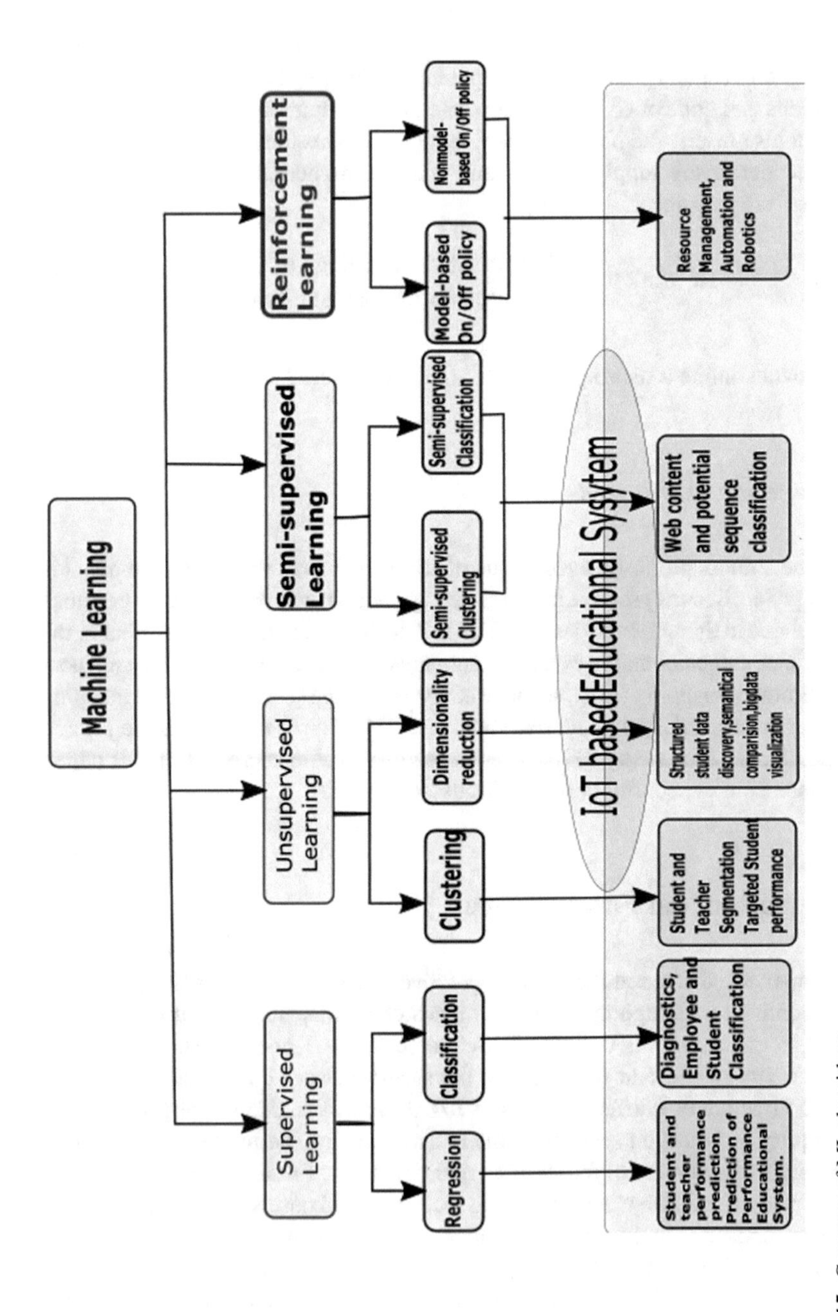

Fig. 15 Category of ML algorithms

which provides security for devices to get rid of unauthorized access of the device and negotiate the devices.

In recent years, there has been rapid advancement in IoT in every sector. It is being seen that most examiners trust in platforms such as intelligent transportation and logistic monitoring. The security challenges identified in the IoT are managed and accomplished to the development. According to a survey, the most problematic thing in IoT is the security layer feeding. The data are being stolen or leaked through layers. The cryptographic algorithm and blockchain algorithm help to secure layer from malicious attack. Most IoT gadgets prefer to have authentication from the server-side. We apply some algorithms like SDN which authenticates the clients during entry such that data packets are secured in the application layer of SDN. We can enhance the security in the IoT environment by implementing various machine learning and deep learning algorithms in different levels of IoT.

References

1. Tripathi, G., Ahad, M. A., & Agarwal, P. (2018). Learning analytics for IoE based educational model using deep learning techniques: Architecture, challenges, and applications. *Smart Learning Environments, 5*(1), 7.
2. Sha, K., Errabelly, R., Wei, W., Yang, T. A., & Wang, Z. (2017). Edgesec: Design of an edge layer security service to enhance IoT security. In *2017 IEEE 1st International Conference on Fog and Edge Computing (ICFEC)* (pp. 81–88). IEEE.
3. Hussain, F., Hussain, R., Hassan, S. A., & Hossain, E. (2019). Machine learning in IoT security: Current solutions and future challenges. arXiv preprint arXiv:1904.05735.
4. Kumari, S., Karuppiah, M., Das, A. K., Li, X., Wu, F., & Kumar, N. (2018). A secure authentication scheme based on elliptic curve cryptography for IoT and cloud servers. *The Journal of Supercomputing, 74*(12), 6428–6453.
5. Tripathi, G., & Ahad, M. A. (2019). IoT in education: An integration of educator community to promote holistic teaching and learning. In *Soft computing in data analytics* (pp. 675–683). Singapore: Springer.
6. Sicato, S., Costa, J., Sharma, P. K., Loia, V., & Park, J. H. (2019). VPNFilter malware analysis on the cyber threats in smart home networks. *Applied Sciences, 9*(13), 2763.
7. Chen, Y., Trappe, W., & Martin, R. P. (2007). Detecting and localizing wireless spoofing attacks. In *2007 4th Annual IEEE Communications Society Conference on Sensor, Mesh and Ad Hoc Communications and Networks* (pp. 193–202). IEEE.
8. Xiao, L., Greenstein, L. J., Mandayam, N. B., & Trappe, W. (2009). Channel-based detection of sybil attacks in wireless networks. *IEEE Transactions on Information Forensics and Security, 4*(3), 492–503.
9. OWASP. (2016). Top IoT vulnerabilities. https://www.owasp.org/index.php/Top_IoT_Vulner abilities.
10. Bhattasali, T., & Chaki, R. (2011). A survey of recent intrusion detection systems for wireless sensor network. In *International Conference on Network Security and Applications* (pp. 268–280). Berlin, Heidelberg: Springer.
11. Kim, H. (2008). Protection against packet fragmentation attacks at 6lowpan adaptation layer. In *2008 International Conference on Convergence and Hybrid Information Technology* (pp. 796–801). IEEE.
12. Hummen, R., Hiller, J., Wirtz, H., Henze, M., Shafagh, H., & Wehrle, K. (2013). 6LoWPAN fragmentation attacks and mitigation mechanisms. In *Proceedings of the Sixth ACM Conference on Security and Privacy in Wireless and Mobile Networks* (pp. 55–66). ACM.

13. Riaz, R., Kim, K. H., & Ahmed, H. F. (2009). Security analysis survey and framework design for IP connected lowpans. In *2009 International Symposium on Autonomous Decentralized Systems* (pp. 1–6). IEEE.
14. Dvir, A., & Buttyan, L. (2011). VeRA-version number and rank authentication in rpl. In *2011 IEEE Eighth International Conference on Mobile Ad-Hoc and Sensor Systems* (pp. 709–714). IEEE.
15. Weekly, K., & Pister, K. (2012). Evaluating sinkhole defense techniques in RPL networks. In *2012 20th IEEE International Conference on Network Protocols (ICNP)* (pp. 1–6). IEEE.
16. Ahmed, F., & Ko, Y. B. (2016). Mitigation of black hole attacks in routing protocol for low power and lossy networks. *Security and Communication Networks, 9*(18), 5143–5154.
17. Pirzada, A. A., & McDonald, C. (2005). Circumventing sinkholes and wormholes in wireless sensor networks. In *IWWAN'05: Proceedings of International Workshop on Wireless Ad-hoc Networks* (Vol. 71).
18. Wang, W., Kong, J., Bhargava, B., & Gerla, M. (2008). Visualisation of wormholes in underwater sensor networks: A distributed approach. *International Journal of Security and Networks, 3*(1), 10–23.
19. Wazid, M., Das, A. K., Kumari, S., & Khan, M. K. (2016). Design of sinkhole node detection mechanism for hierarchical wireless sensor networks. *Security and Communication Networks,9*(17), 4596–4614.
20. Zhang, K., Liang, X., Lu, R., & Shen, X. (2014). Sybil attacks and their defenses in the internet of things. *IEEE Internet of Things Journal, 1*(5), 372–383.
21. Konolige, T., Wilson, C., Wang, G., Wang, X., Zheng, H., & Zhao, B. Y. (2013). You are how you click: Clickstream analysis for sybil detection. In *Presented as part of the 22nd {USENIX} Security Symposium ({USENIX} Security* 13) (pp. 241–256).
22. Granjal, J., Monteiro, E., & Silva, J. S. (2014). Network layer security for the internet of things using TinyOS and BLIP. *International Journal of Communication Systems, 27*(10), 1938–1963.
23. Raza, S., Duquennoy, S., Voigt, T., & Roedig, U. (2011). Demo abstract: Securing communication in 6LoWPAN with compressed IPsec. In *2011 International Conference on Distributed Computing in Sensor Systems and Workshops* (*DCOSS*) (pp. 1–2). IEEE.
24. Granjal, J., Monteiro, E., & Silva, J. S. (2010). Enabling network-layer security on IPv6 wireless sensor networks. In *2010 IEEE Global Telecommunications Conference GLOBECOM 2010* (pp. 1–6). IEEE.
25. Mahalle, P. N., Anggorojati, B., Prasad, N. R., & Prasad, R. (2013). Identity authentication and capability-based access control (ICAC) for the internet of things. *Journal of Cyber Security and Mobility, 1*(4), 309–348.
26. Mahalle, P. N., Anggorojati, B., Prasad, N. R., & Prasad, R. (2013). Identity authentication and capability-based access control (iacac) for the internet of things. *Journal of Cyber Security and Mobility, 1*(4), 309–348.
27. Brachmann, M., Garcia-Morchon, O., & Kirsche, M. (2011). Security for practical coap applications: Issues and solution approaches. *GI/ITG KuVSFachgesprchSensornetze (FGSN). Universitt Stuttgart.*
28. Granjal, J., Monteiro, E., & Silva, J. S. (2013). End-to-end transport-layer security for internet-integrated sensing applications with mutual and delegated ECC public-key authentication. In *2013 IFIP Networking Conference* (pp. 1–9). IEEE.
29. Peretti, G., Lakkundi, V., & Zorzi, M. (2015). BlinkToSCoAP: An end-to-end security framework for the internet of things. In *2015 7th International Conference on Communication Systems and Networks (COMSNETS)* (pp. 1–6). IEEE.
30. Raza, S., Voigt, T., & Jutvik, V. (2012). Lightweight IKEv2: A key management solution for both the compressed IPsec and the IEEE 802.15. 4 security. In *Proceedings of the IETF Workshop on Smart Object Security* (Vol. 23).
31. Henze, M., Wolters, B., Matzutt, R., Zimmermann, T., & Wehrle, K. (2017). Distributed configuration, authorization, and management in the cloud-based internet of things. In *2017 IEEE Trustcom/BigDataSE/ICESS* (pp. 185–192). IEEE.

32. Brachmann, M., Keoh, S. L., Morchon, O. G., & Kumar, S. S. (2012). End-to-end transport security in the IP-based Internet of things. In *2012 21st International Conference on Computer Communications and Networks (ICCCN)* (pp. 1–5). IEEE.
33. Granjal, J., Monteiro, E., & Silva, J. S. (2013). Application-layer security for the WoT: Extending CoAP to support end-to-end message security for internet-integrated sensing applications. In *International Conference on Wired/Wireless Internet Communication* (pp. 140–153). Berlin, Heidelberg: Springer.
34. Sethi, M., Arkko, J., & Keränen, A. (2012). End-to-end security for sleepy smart object networks. In *37th Annual IEEE Conference on Local Computer Networks-Workshops* (pp. 964–972). IEEE.
35. Conzon, D., Bolognesi, T., Brizzi, P., Lotito, A., Tomasi, R., & Spirito, M. A. (2012). The virus middleware: An XMPP based architecture for secure IoT communications. In *2012 21st International Conference on Computer Communications and Networks (ICCCN)* (pp. 1–6). IEEE.
36. Liu, C. H., Yang, B., & Liu, T. (2014). Efficient naming, addressing and profile services in internet-of-things sensory environments. *Ad Hoc Networks, 18*, 85–101.
37. Gul, S., Asif, M., Ahmad, S., Yasir, M., Majid, M., & Arshad, M. S. (2017). A survey on the role of the internet of things in education. *IJCSNS, 17*(5), 159.
38. Healion, D., Russell, S., Cukurova, M., & Spikol, D. (2017). Tracing physical movement during practice-based learning through multimodal learning analytics. In *ACM International Conference Proceeding Series* (Vol. 7, pp. 588–589). *Seventh International Learning Analytics and Knowledge Conference (LAK'17)*. Association for Computing Machinery (ACM), Vancouver, BC, Canada.
39. Smith, A., Min, W., Mott, B. W., & Lester, J. C. (2015). Diagrammatic student models: Modeling student drawing performance with deep learning. In *International Conference on User Modeling, Adaptation, and Personalization* (pp. 216–227). Cham: Springer.
40. Park, N., & Kang, N. (2016). Mutual authentication scheme in secure internet of things technology for comfortable lifestyle. *Sensors, 16*(1), 20.
41. Robles, R. J., & Endencio-Robles, D. (2019). State of internet of things (IoT) security attacks, vulnerabilities and solutions. *Computer Reviews Journal, 3*, 255–263.
42. Fidalgo, A., Sein-Echaluce, M., Conde, M. A., & Garcia-Peñalvo, F. J. (2014). Design and development of a learning analytics system to evaluate group work competence. In *2014 9th Iberian Conference on Information Systems and Technologies (CISTI)* (pp. 1–6). IEEE.
43. Bagheri, M., & Movahed, S. H. (2016). The effect of the internet of things (IoT) on education business model. In *12th International Conference on Signal-Image Technology and Internet-Based Systems (SITIS)* (pp. 435–441). IEEE, Naples, Italy.
44. Atherton, M., Shah, M., Vazquez, J., Griffiths, Z., Jackson, B., & Burgess, C. (2017). Using learning analytics to assess student engagement and academic outcomes in open access enabling programs. *Open Learning: The Journal of Open, Distance and e-Learning, 32*(2), 119–136.
45. Zhang, S., Li, Z., Zhang, Y., Chen, K., Deng, Q., Ray, S., & Jin, Y. (2018). Internet-of-things security and vulnerabilities: Taxonomy, challenges, and practice. *Journal of Hardware and Systems Security, 2*(2), 97–110.
46. Zhou, J., Cao, Z., Dong, X., & Vasilakos, A. V. (2017). Security and privacy for cloud-based IoT: Challenges. *IEEE Communications Magazine, 55*(1), 26–33.

Author Index

R. Kumar et al. (eds.), *Multimedia Technologies in the Internet of Things Environment*, Studies in Big Data 79, https://doi.org/10.1007/978-981-15-7965-3

Printed in the United States
by Baker & Taylor Publisher Services